Springer Series in Statistics

Leo A. Goodman
William H. Kruskal

Measures of Association for Cross Classifications

Springer-Verlag
New York Heidelberg Berlin

Leo A. Goodman
Department of Statistics
University of Chicago
Chicago, Illinois 60637
USA

William H. Kruskal
Department of Statistics
University of Chicago
Chicago, Illinois 60637
USA

Library of Congress Cataloging in Publication Data

Goodman, Leo A
Measures of association for cross classifications.

(Springer series in statistics; v. 1)
Includes bibliographies.
1. Sociology—Methodology. 2. Sociology—Statistical methods. I. Kruskal, William, 1919- joint author. III. Title. III. Title: Cross classifications. IV. Series.
HM24.G627 301'.01'82 79-19570

Printed in the United States of America.

9 8 7 6 5 4 3 2 1

ISBN 0-387-90443-3 Springer-Verlag New York
ISBN 3-540-90443-3 Springer-Verlag Berlin Heidelberg

Foreword

In 1954, prior to the era of modern high speed computers, Leo A. Goodman and William H. Kruskal published the first of a series of four landmark papers on measures of association for cross classifications. By describing each of several cross classifications using one or more *interpretable* measures, they aimed to guide other investigators in the use of sensible data summaries. Because of their clarity of exposition, and their thoughtful statistical approach to such a complex problem, the guidance in this paper is as useful and important today as it was on its publication 25 years ago.

Summarizing association in a cross-classification by a single number inevitably loses information. Only by the thoughtful choice of a measure of association can one hope to lose only the less important information and thus arrive at a satisfactory data summary. The series of four papers reprinted here serve as an outstanding guide to the choice of such measures and their use.

Many users view measures of association as they do correlations, applicable to essentially all data sets. To their credit, Goodman and Kruskal argue that ideally each research problem should have one or possibly several measures of association, with operational meaning, developed for its unique needs. Because the Goodman-Kruskal papers provide what amounts to a comprehensive catalogue of existing measures (several of which they themselves created), analysts may begin by examining and attempting to choose wisely from those measures currently available. If none are satisfactory, and new ones are created, the Goodman-Kruskal papers will be helpful as models and guides.

This series of papers evolved over a twenty year period. The first and core paper appeared in 1954. It suggests criteria for judging measures of association and introduces several new measures with specific contextual meanings. Examples and illustrations abound. The 1959 paper serves as a supplement to the inital one and provides additional historical and bibliographic material. The 1963 paper derives large-sample standard errors for the sample analogues of population measures of association and presents some numerical results about the adequacy of large-sample normal approximations. The 1972 paper presents a new look at the asymptotics, and provides a more unified way to derive large-sample variances for those measures of association that can be expressed as ratios of functions of the cell probabilities. Thus the techniques can be used for tried and true measures, and also for ones not yet invented. Only by rereading these papers many times can one appreciate the perspicacity that the authors have brought to this perplexing problem.

As a colleague of Leo and Bill at The University of Chicago, I was privileged to witness the care and scholarly attention they gave to the last of the measures of association papers. It was truly a labor of love. Thus I am delighted both personally and as a member of the Editorial Advisory Board for the Springer Statistical Series that Springer-Verlag has been able to bring together these four papers in a single volume, so that they can be shared with a new generation of statisticians and scientists.

August, 1979 STEPHEN E. FIENBERG

Preface*

In the early 1950s, as young faculty members at the University of Chicago, we had separate conversations with senior colleagues there about statistical treatment of data that were naturally arranged as cross classifications of counts. One of us talked to Bernard Berelson (then Dean of the Graduate Library School and later the President of the Population Council), who was at that time dealing with extensive cross classifications related to voting behavior. For example, he might have a number of cross classifications of intended vote against educational level for different sections of a city.

The other conversations were with the late Louis Thurstone (a major figure in the field of psychometrics, and in particular in the development of factor analysis), who also was dealing with multiple cross classifications in the context of the relationships between various personal characteristics (e.g., leadership ability) and results from various psychological tests.

In both cases the investigator had substantial numbers of cross classifications and needed a sensible way to reduce the data to try to make it coherent. One promising approach was felt to be replacement of each cross classification by a number (or numbers) that measured in a reasonable way the degree of association between the characteristics corresponding to the rows and columns of the tabulated cross classification.

Thus, the two of us were independently thinking about the same question. We discovered our mutual interest during a conversation at a party—we think that it was a New Year's Eve party at the Quadrangle (Faculty) Club—and the paper grew out of that interaction.

We knew something of the existing literature on measures of association for

*Most of this preface appeared in "This Week's Citation Classic", *Current Contents,* Social and Behavioral Sciences, No. 26, 25 June 1979, page 14.

cross classifications, and as we studied it further we recognized that most suggested measures of association were formal and arbitrary, without relevant interpretations—or without interpretations at all. Our contribution was to suggest a number of association measures that have interesting interpretations and to provide a simple taxonomy for cross classifications. As an example of the latter, we emphasized the importance of knowing whether or not the categories of a classification have not a natural ordering.

Since cross classifications occur throughout science, since our emphasis on interpretation was perhaps novel, and since our work was quickly incorporated into textbook expositions, citations to the paper became numerous. We continued work on the topic, digging more deeply into its history and fields of application, and treating at length the relevant approximate sampling theory in an effort to contribute some new approaches and to effect some changes in statistical thinking and practice.

One of us also developed an interest in ordinal measures of association beyond cross classifications as such.[1] The other was led to extensive research in the analysis of multi-way cross classifications, leading to what have come to be known as log-linear model theory and methodology.[2] Another outgrowth, we dare to hope, of our paper has been fresh general concern with descriptive statistics from the viewpoint of finding usefully interpretable characteristics of populations and samples.

In this reprinting, notes appear in the margin at a few points to indicate errors that were corrected in later papers of the sequence. One additional trivial error has been directly corrected. Otherwise the papers appear just as they originally appeared.

We end this preface with a statement of thanks to W. Allan Wallis, first Chairman of the Department of Statistics at the University of Chicago. There are many reasons for us to thank him, but the relevant one now is that he introduced us to Berelson and to Thurstone, and from those introductions our thinking on measures of association arose. Wallis, in fact, did far more than perform introduction: he discussed our nascent work with us, and suggested an important approach with which his name is associated in our first paper.

Chicago, Illinois
September, 1979

Leo A. Goodman
William H. Kruskal

[1]Kruskal, W. H. Ordinal measures of association. *J. Amer. Statist. Assoc.* 53:814–61, 1958.

[2]Goodman, L. A. The multivariate analysis of qualitative data: interactions among multiple classifications. *J. Amer. Statist. Assoc.* 65:226–56, 1970.

Contents

Measures of Association for Cross Classifications 2

1. Introduction 3
2. Four Preliminary Considerations 5
3. Conventions 8
4. Traditional Measures 9
5. Measures Based on Optimal Prediction 10
6. Measures Based upon Optimal Prediction of Order 17
7. The Generation of Measures by the Introduction of Loss Functions 24
8. Reliability Models 26
9. Proportional Prediction 29
10. Association with a Particular Category 30
11. Partial Association 30
12. Multiple Association 31
13. Sampling Problems 32
14. Concluding Remarks 33
15. References 33

Measures of Association for Cross Classifications. II: Further Discussion and References 35

1. Introduction and Summary 35
2. Supplementary Discussion to Prior Paper 37
3. Work on Measures of Association in the Late Nineteenth and Early Twentieth Centuries 39
4. More Recent Publications 49
5. References 68

Measures of Association for Cross Classifications. III: Approximate Sampling Theory 76

1. Introduction and Summary 77
2. Notation and Preliminaries 79
3. Multinominal Sampling over the Whole Double Polytomy 81
4. Multinomial Sampling within Each Row (Column) of the Double Polytomy 114
5. Further Remarks 120
6. References 121
   Appendix 122

Measures of Association for Cross Classifications. IV: Simplification of Asymptotic Variances 131

1. Introduction and Summary 131
2. Multinomial Sampling over the Entire Two-Way Cross Classification 132
3. Independent Multinomial Sampling in the Rows 137
4. Use of the Results in Practice 141
5. When Does $\sigma=0$? 142
6. Cautionary Note about Asymptotic Variances 145
   References 146

Measures of Association for Cross Classifications

Reprinted from the JOURNAL OF THE AMERICAN STATISTICAL ASSOCIATION
December, 1954, Vol. 49, pp. 732-764

MEASURES OF ASSOCIATION FOR CROSS CLASSIFICATIONS*

LEO A. GOODMAN AND WILLIAM H. KRUSKAL
University of Chicago

CONTENTS

	Page
1. INTRODUCTION	733
2. FOUR PRELIMINARY CONSIDERATIONS	735
2.1. Continua	735
2.2. Order	736
2.3. Symmetry	736
2.4. Manner of Formation of the Classes	737
3. CONVENTIONS	738
4. TRADITIONAL MEASURES	739
5. MEASURES BASED ON OPTIMAL PREDICTION	740
5.1. Asymmetrical Optimal Prediction. A Particular Model of Activity	740
5.2. Symmetrical Optimal Prediction. Another Model of Activity	742
5.3. An Example	744
5.4. Weighting Columns or Rows	745
6. MEASURES BASED UPON OPTIMAL PREDICTION OF ORDER	747
6.1. Preliminaries	747
6.2. A Proposed Measure	748
6.3. An Example	751
7. THE GENERATION OF MEASURES BY THE INTRODUCTION OF LOSS FUNCTIONS	754
7.1. Models Based on Loss Functions	754
7.2. Loss Functions and the λ Measures	755
7.3. The Conventional Measures in Terms of Loss Functions	755
8. RELIABILITY MODELS	756
8.1. Generalities	756
8.2. A Measure of Reliability in the Unordered Case	757
8.3. Reliability in the Ordered Case	758
9. PROPORTIONAL PREDICTION	759
10. ASSOCIATION WITH A PARTICULAR CATEGORY	760
11. PARTIAL ASSOCIATION	760
11.1. Simple Average of λ_b	761
11.2. Measure Based Directly on Probabilities of Error	761
12. MULTIPLE ASSOCIATION	761
13. SAMPLING PROBLEMS	762
14. CONCLUDING REMARKS	763
15. REFERENCES	763

* This paper is partly an outgrowth of work sponsored by the Army, Navy, and Air Force through the Joint Services Advisory Committee for Research Groups in Applied Mathematics and Statistics by Contract No. N6ori-02035. We are indebted for helpful comments and criticisms to: Otis D. Duncan (University of Chicago), Churchill Eisenhart (National Bureau of Standards), Maurice G. Kendall

When populations are cross-classified with respect to two or more classifications or polytomies, questions often arise about the degree of association existing between the several polytomies. Most of the traditional measures or indices of association are based upon the standard chi-square statistic or on an assumption of underlying joint normality. In this paper a number of alternative measures are considered, almost all based upon a probabilistic model for activity to which the cross-classification may typically lead. Only the case in which the population is completely known is considered, so no question of sampling or measurement error appears. We hope, however, to publish before long some approximate distributions for sample estimators of the measures we propose, and approximate tests of hypotheses. Our major theme is that the measures of association used by an empirical investigator should not be blindly chosen because of tradition and convention only, although these factors may properly be given some weight, but should be constructed in a manner having operational meaning within the context of the particular problem.

1. INTRODUCTION

Many studies, particularly in the social sciences, deal with populations of individuals which are thought of as cross-classified by two or more polytomies. For example, the adult individuals living in New York City may be classified as to

Borough:	5 classes
Newspaper most often read:	perhaps 6 classes
Television set in home or not:	2 classes
Level of formal education:	perhaps 5 classes
Age:	perhaps 10 classes

For simplicity we deal largely with the case of two polytomies, although many of our remarks may be extended to a greater number. The double polytomy is the most common, no doubt because of the ease with which it can be tabulated and displayed on the printed page. Most of our remarks suppose the population completely known in regard to the classifications, and indeed this seems to be the way to begin in the construction of rational measures of association. After agreement has been reached on the utility of a measure for a known population, then

(London School of Economics and Political Science), Frederick Mosteller (Harvard University), I. Richard Savage (National Bureau of Standards), Alan Stuart (London School of Economics and Political Science), Louis L. Thurstone (University of North Carolina), John W. Tukey (Princeton (University), W. Allen Wallis (University of Chicago), and E. J. Williams (Commonwealth Scientific and Industrial Research Organization, Australia). Part of Mr. Goodman's work on this paper was carried out at the Statistical Laboratory of the University of Cambridge under a Fulbright Award and a Social Science Research Council Fellowship. The authors were led to work on the problems of this paper as a result of conversations with Louis L. Thurstone and Bernard R. Berelson.

one should consider the sampling problems associated with estimation and tests about this population parameter.

A double polytomy may be represented by a table of the following kind:[1]

A	B				
	B_1	B_2	$\cdots$	B_β	Total
A_1	ρ_{11}	ρ_{12}	$\cdots$	$\rho_{1\beta}$	$\rho_{1\cdot}$
A_2	ρ_{21}	ρ_{22}	$\cdots$	$\rho_{2\beta}$	$\rho_{2\cdot}$
$\vdots$	$\vdots$	$\vdots$	$\vdots$	$\vdots$	$\vdots$
A_α	$\rho_{\alpha 1}$	$\rho_{\alpha 2}$	$\cdots$	$\rho_{\alpha\beta}$	$\rho_{\alpha\cdot}$
Total	$\rho_{\cdot 1}$	$\rho_{\cdot 2}$	$\cdots$	$\rho_{\cdot\beta}$	1

where

Classification A divides the population into the α classes $A_1, A_2, \cdots, A_\alpha$.

Classification B divides the population into the β classes $B_1, B_2, \cdots, B_\beta$.

The proportion of the population that is classified as both A_a and B_b is ρ_{ab}.

The marginal proportions will be denoted by

$\rho_{a\cdot}$ = the proportion of the population classified as A_a.

$\rho_{\cdot b}$ = the proportion of the population classified as B_b.

If the use to which a measure of association were to be put could be precisely stated, there would be little difficulty in defining an appropriate measure. For example, using the above cross-classification of the New York City population, a television service company might wish to

[1] Tables of this kind are frequently called *contingency tables*. We shall not use this term because of its connotation of a specific sampling scheme when the population is not known and one infers on the basis of a sample.

place a single newspaper advertisement which would be read by as many prospective customers as possible. Then the important information from the table of newspaper-most-often-read vs. television-set-in-home-or-not would be: which newspaper is most often read among those with television sets? And a reasonable measure of association would simply be the proportion of those with television sets who read this newspaper.

It is rarely the case, however, that the purpose of an investigation can be so specifically stated. More typically an investigation is exploratory or has a multiplicity of goals. Sometimes a measure of association is desired simply so that a large mass of data may be summarized compactly.

The basic theme of this paper is that, even though a single precise goal for an investigation cannot be specified, it is still possible and desirable to choose a measure of association which has contextual meaning, instead of using as a matter of course one of the traditional measures. In order to choose a measure of association which has meaning we propose the construction of probabilistic models of predictive activity, the particular model to be chosen in the light of the particular investigation at hand. The measure of association will then be a probability, or perhaps some simple function of probabilities, within such a model. Such is our general contention; most of the remainder of this paper is concerned with its exemplification in particular instances.

We wish to emphasize that the specific measures of association described here are *not* presented as factotum or universal measures. Rather, they are suggested as reasonable for use in appropriate circumstances only, and even in those circumstances other measures may and should be considered and investigated.

A good deal of attention has been paid in the literature to the special case of two dichotomies. We are more interested here in measures of association suitable for use with any numbers of classes in the polytomies or classifications.

2. FOUR PRELIMINARY CONSIDERATIONS

Four distinctions or cautionary remarks should be made early in any discussion of measures of association.

2.1. *Continua*

We may or may not wish to think of a polytomy as arising from an underlying continuum. For example, age may for convenience be divided into ten classifications, but it clearly does arise from an underlying continuum; however, newspaper-most-often-read would scarcely be so construed. If a polytomy does arise from an underlying continuum one may or may not wish to assume that the population has some specific kind of distribution with respect to it.

In those cases in which all the polytomies of a study arise jointly from a multivariate normal distribution on an underlying continuum, one would naturally turn to measures of association based on the correlation coefficients. These in turn might well be estimated from a sample by the tetrachoric correlation coefficient method or a generalization of it. In some cases one polytomy may arise from a continuum and the other not. An interesting discussion of this case for two dichotomies was given in 1915 by Greenwood and Yule ([3], Section 3). We do not discuss either of these cases in this paper, but restrict ourselves to situations in which there are no relevant underlying continua.

The desirability of assuming an underlying joint continuum was one of the issues of a heated debate forty years ago between Yule [15] on the one hand and K. Pearson and Heron [9] on the other. Yule's position was that very frequently it is misleading and artificial to assume underlying continua; Pearson and Heron argued that almost always such an assumption is both justified and fruitful.

2.2. *Order*

There may or may not be an underlying order between the classifications of a polytomy. For example "level of formal education" admits an obvious ordering; but borough of residence would not usually be thought of in an ordered way. If there is an ordering, it may or may not be relevant to the investigation. Sometimes an ordering may be important but not its direction. If there is an underlying one-dimensional continuum, it establishes an ordering.

When there is no natural or relevant ordering of the classes of a polytomy, one may reasonably ask that a measure of association not depend on the particular order in which the classes are tabulated.

2.3. *Symmetry*

It may or may not be that one looks at two polytomies symmetrically. When we are sure a priori that a causal relationship (if it exists) runs in one direction but not the other, then our viewpoint will be asymmetric. This will also happen if one plans to *use* the results of the experiment in one direction only. On the other hand, there is often no reason to give one polytomy precedence over another.

2.4. *Manner of Formation of the Classes*

Decisions about the definitions of the classes of a polytomy, or changes from a finer to a coarser classification (or vice-versa), can affect all the measures of association of which we know. For example, suppose we begin with the 4×4 table

0	.25	0	0
.25	0	0	0
0	0	0	.25
0	0	.25	0

and combine neighboring pairs of classes. We obtain

.5	0
0	.5

which might greatly change a measure of association. Or we might combine the three bottom rows and the three right-hand columns. This gives

0	.25
.25	.5

which presents quite a different intuitive degree of association. By other poolings one can obtain other 2×2 tables.

Although this example is extreme, similar changes can be made in the character of almost any cross-classification table. Related examples are discussed by Yule [15].

At first this consideration might seem to vitiate any reasonable discussion of measures of association. We feel, however, that it is in fact desirable that a measure of association reflect the classes *as defined for the data.* Thus one should not speak, for example, of association between income level and level of formal education without specifying particular class definitions. Of course, in many cases association—however measured—would not be much affected by any reasonable redefinition of the classes, and then the above finicky form of statement can be simplified. That the definition of the classes can affect the degree of association naturally means that careful attention should be given to the class definitions in the light of the expected uses of the final conclusions.

3. CONVENTIONS

It is conventional, and often convenient, to set up a measure of association so that either

(*i*) It takes values between -1 and $+1$ inclusive, is -1 or $+1$ in case of "complete association," and is zero in the case of independence.

(*ii*) It takes values between 0 and $+1$ inclusive, is $+1$ in the case of "complete association," and is zero in the case of independence.

Convention (*i*) is appropriate when the association is thought of as signed (e.g., association between income and dollars spent is positive, between income and per cent of income spent is negative). Convention (*ii*) is appropriate when no such sign considerations exist, as when there is no natural order.

"Complete association," as we shall see, is somewhat ambiguous. "Independence," on the other hand, has its usual meaning, that is

$$\rho_{ab} = \rho_{a\cdot}\rho_{\cdot b} \quad (a = 1, \cdots, \alpha;\ b = 1, \cdots, \beta). \tag{1}$$

Conventions like these have seemed important to some authors, but we believe they diminish in importance as the meaningfulness of the measure of association increases. One real danger connected with such conventions is that the investigator may carry over size preconceptions based upon experience with completely different measures subject to the same conventions. For example, some elementary statistics textbooks warn that a population correlation coefficient less than about .5 in absolute value may have little practical significance, in the sense that then the conditional variance is not much less than the marginal variance. Research workers in various fields thus tend to develop rather strong feelings that population correlation coefficients less than, say, .5, have little substantive importance. The same feelings might be carried over, without justification, to all other measures of association so defined as to lie between +1 and −1.

It should also be mentioned that once one has a measure of association satisfying one of the above conventions, then an infinite number of others also satisfying the same convention can be obtained—for example, by raising to a power and adjusting the sign if necessary.

4. TRADITIONAL MEASURES

Excellent accounts of these may be found in [16], Chaps. 2 and 3, and [7], Chap. 13. Many of these stem from the standard chi-square statistic upon which a test of independence is usually based. If a finite population has ν members and we set $\nu_{ab}=\nu\rho_{ab}$, $\nu_{a\cdot}=\nu\rho_{a\cdot}$, $\nu_{\cdot b}=\nu\rho_{\cdot b}$, etc., the chi-square statistic in the case of two classifications is

$$(2) \qquad \chi^2 = \sum_a \sum_b \frac{(\nu_{ab} - \nu_{a\cdot}\nu_{\cdot b}/\nu)^2}{\nu_{a\cdot}\nu_{\cdot b}/\nu} = \nu \sum_a \sum_b \frac{(\rho_{ab} - \rho_{a\cdot}\rho_{\cdot b})^2}{\rho_{a\cdot}\rho_{\cdot b}}$$

$$= \nu \sum_a \sum_b \frac{\rho_{ab}{}^2}{\rho_{a\cdot}\rho_{\cdot b}} - \nu.$$

A great deal of attention has been given to the case $\alpha=\beta=2$. For this special case Yule has defined the following coefficient of association:

$$(3) \qquad Q = \frac{\nu_{11}\nu_{22} - \nu_{12}\nu_{21}}{\nu_{11}\nu_{22} + \nu_{12}\nu_{21}}$$

whose numerator squared is essentially the same as that of a convenient and popular form for χ^2 in the 2×2 case. Another coefficient suggested by Yule for the 2×2 case is

$$(4) \qquad Y = \frac{\sqrt{\nu_{11}\nu_{22}} - \sqrt{\nu_{12}\nu_{21}}}{\sqrt{\nu_{11}\nu_{22}} + \sqrt{\nu_{12}\nu_{21}}}\,.$$

A coefficient often used for the general $\alpha\times\beta$ case is simply χ^2/ν, often called the mean square contingency and denoted by ϕ^2. A variation of this, suggested by Karl Pearson, is

$$(5) \qquad C = \sqrt{\frac{[\chi^2/\nu]}{1 + \chi^2/\nu}}$$

which has been called the coefficient of contingency, or the coefficient of mean square contingency. Another variation, proposed by Tschuprow, is

* (6) $$T = \sqrt{[\chi^2/\nu]/(\alpha - 1)(\beta - 1)}.$$

The last two suggestions, according to Kendall [7], were made in attempts to norm χ^2 so that it might lie between 0 and 1 and take the extreme values under independence and "complete association." Cramér ([1], p. 282) suggests the following variant:

(7) $$[\chi^2/\nu]/\text{Min}\ (\alpha - 1, \beta - 1)$$

which gives a better norming than does C or T since it lies between 0 and 1 and actually attains both end points appropriately. Cramér's suggestion does not seem to be well known by workers using this general kind of index.

The fact that an excellent test of independence may be based on χ^2 does not at all mean that χ^2, or some simple function of it, is an appropriate *measure* of degree of association. A discussion of this point is presented by R. A. Fisher ([2], Section 21). We have been unable to find any convincing published defense of χ^2-like statistics as measures of association.

One difficulty with the use of the traditional measures, or of any measures that are not given operational interpretation, is that it is difficult to compare meaningfully their values for two cross-classifications. Suppose that C turns out to be .56 and .24 respectively in two cross-classification tables. One wants to be able to say that there is higher association in the first table than the second, but investigators sometimes restrain themselves, with commendable caution, from making such a comparison. Their restraint may stem in part from the noninterpretability of C. (Of course, when samples are small they may also be restrained by inadequate knowledge of sampling fluctuation.)

One class of measures that will not be discussed here is characterized by the assignment of numerical scores to the classes, followed by the use of the correlation coefficient on these scores. A recent article on such measures is by E. J. Williams [12]. It contains references leading back to earlier literature. We feel that the use of arbitrary scores to motivate measures is infrequently appropriate, but it should be pointed out that measures not motivated by the correlation of scores can often be thought of from the score viewpoint.

5. MEASURES BASED ON OPTIMAL PREDICTION

5.1. *Asymmetrical Optimal Prediction. A Particular Model of Activity*

Let us consider first a probabilistic model which might be useful in a situation of the following kind:

(*i*) Two polytomies, A and B.
(*ii*) No relevant underlying continua.
(*iii*) No natural ordering of interest.
(*iv*) Asymmetry holds: The A classification precedes the B classification chronologically, causally, or otherwise.

An example of such a situation might be a study of the association between college attended (A) and kind of adult occupation (B). Our model of activity is the following: An individual is chosen at random from the population and we are asked to guess his B-class as well as we can, either

1. Given no further information, or
2. Given his A class.

Clearly we can do no worse in case 2 than in case 1. Represent by $\rho_{\cdot m}$ the largest marginal proportion among the B classes and by ρ_{am} the largest proportion in the ath row of the cross-classification table—that is

$$(8) \qquad \rho_{\cdot m} = \operatorname*{Max}_b \rho_{\cdot b}, \quad \rho_{am} = \operatorname*{Max}_b \rho_{ab} \cdot$$

Then in case 1 we are best off guessing that B_b for which $\rho_{\cdot b} = \rho_{\cdot m}$—that is, guessing that B class which has the largest marginal proportion—and our probability of error is $1 - \rho_{\cdot m}$. In case 2 we are best off guessing that B_b for which $\rho_{ab} = \rho_{am}$ (letting A_a be the given A class)—that is, guessing that B class that has the largest proportion in the observed A class—and our probability of error is[2] $1 - \sum_a \rho_{am}$.

Then we propose as a measure of association (following Guttman [4])

$$(9) \qquad \lambda_b = \frac{(\text{Prob. of error in case 1}) - (\text{Prob. of error in case 2})}{(\text{Prob. of error in case 1})}$$

$$= \frac{\sum_a \rho_{am} - \rho_{\cdot m}}{1 - \rho_{\cdot m}},$$

which is the relative decrease in probability of error in guessing B_b as between A_a unknown and A_a known. To put this another way, λ_b gives the proportion of errors that can be eliminated by taking account of knowledge of the A classifications of individuals.

Some important properties of λ_b follow:

---

[2] It may be that in case 1 there is more than one b for which $\rho_{\cdot b} = \rho_{\cdot m}$. Then any method of choosing which of these b's to guess—including flipping an appropriately multi-sided die—gives rise to the same probability of error, $1 - \rho_{\cdot m}$. A similar comment applies to case 2.

(*i*) λ_b is indeterminate if and only if the population lies in one column, that is, lies in one B class.

(*ii*) Otherwise the value of λ_b is between 0 and 1 inclusive.

(*iii*) λ_b is 0 if and only if knowledge of the A classification is of no help in predicting the B classification, i.e., if there exists a b_0 such that $\rho_{ab_0} = \rho_{am}$ for all a.

(*iv*) λ_b is 1 if and only if knowledge of an individual's A class completely specifies his B class, i.e., if each row of the cross-classification table contains at most one nonzero ρ_{ab}.

(*v*) In the case of statistical independence λ_b, when determinate, is zero. The converse need not hold: λ_b may be zero without statistical independence holding.

(*vi*) λ_b is unchanged by permutation of rows or columns.

That λ_b may be zero without statistical independence holding may be considered by some as a disadvantage of this measure. We feel, however, that this is not the case, for λ_b is constructed specifically to measure association in a restricted but definite sense, namely the predictive interpretation given. If there is no association in that sense, even though there is association in other senses, one would want λ_b to be zero. Moreover, all the measures of association of which we know are subject to this kind of criticism in one form or another, and indeed it seems inevitable. To obtain a measure of association one must sharpen the definition of association, and this means that of the many vague intuitive notions of the concept some must be dropped.

We may similarly define

$$\lambda_a = \frac{\sum_b \rho_{mb} - \rho_{m\cdot}}{1 - \rho_{m\cdot}}, \tag{10}$$

where

$$\rho_{m\cdot} = \operatorname*{Max}_a \rho_{a\cdot}$$
$$\rho_{mb} = \operatorname*{Max}_a \rho_{ab}. \tag{11}$$

Thus λ_a is the relative decrease in probability of error in guessing A_a as between B_b unknown and known.

So far as we know, λ_a and λ_b were first suggested by Guttman ([4], Part I, 4), and our development of them is very similar to his.

5.2. *Symmetrical Optimal Prediction. Another Model of Activity*

In many cases the situation is symmetrical, and one may alter the model of activity as follows: an individual is chosen at random from the population and we are asked to guess his A class half the time (at random) and his B class half the time (at random) either given:

1. No further information, or
2. The class of the individual other than the one being guessed; that is the individual's A_a when we guess B_b and vice versa.

In case 1 the probability of error is $1-\frac{1}{2}(\rho_{\cdot m}+\rho_{m\cdot})$, and in case 2 the probability of error is $1-\frac{1}{2}(\sum_a \rho_{am}+\sum_b \rho_{mb})$. Hence we may consider the relative decrease in probability of error as we go from case 1 to case 2, and define the coefficient

$$\lambda = \frac{\frac{1}{2}[\sum_a \rho_{am} + \sum_b \rho_{mb} - \rho_{\cdot m} - \rho_{m\cdot}]}{1 - \frac{1}{2}(\rho_{\cdot m} + \rho_{m\cdot})}. \tag{12}$$

Some properties of λ follow:

(*i*) λ is determinate except when the entire population lies in a single cell of the table.

(*ii*) Otherwise the value of λ is between 0 and 1 inclusive.

(*iii*) λ is 1 if and only if all the population is concentrated in cells no two of which are in the same row or column.

(*iv*) λ is 0 in the case of statistical independence, but the converse need not hold.

(*v*) λ is unchanged by permutations of rows or columns.

(*vi*) λ lies between λ_a and λ_b inclusive.

The computation of λ_a, λ_b, or λ is extremely simple. Usually one is given the population, not in terms of the ρ_{ab}'s but rather in terms of the numbers of individuals in each cell. Let ν be the total number of individuals in the population, $\nu_{ab}=\nu\rho_{ab}$, $\nu_{am}=\nu\rho_{am}$, $\nu_{mb}=\nu\rho_{mb}$, and so on. Then

$$\lambda_b = \frac{\sum_a \nu_{am} - \nu_{\cdot m}}{\nu - \nu_{\cdot m}}, \tag{13}$$

$$\lambda_a = \frac{\sum_b \nu_{mb} - \nu_{m\cdot}}{\nu - \nu_{m\cdot}}, \tag{14}$$

$$\lambda = \frac{\sum_a \nu_{am} + \sum_b \nu_{mb} - \nu_{\cdot m} - \nu_{m\cdot}}{2\nu - (\nu_{\cdot m} + \nu_{m\cdot})}. \tag{15}$$

5.3. *An example*

The following table is taken from reference [7], p. 300, and originally was given by Ammon in "Zur Anthropologie der Badener." It deals with hair and eye color of males. The table is given in terms of the ν_{ab}'s. A_1, A_2, A_3 are respectively Blue, Grey or Green, Brown; B_1, B_2, B_3, B_4 are respectively Fair, Brown, Black, Red.

Eye Color Group	Hair Color Group				
	B_1	B_2	B_3	B_4	$\nu_{a\cdot}$
A_1	1768	807	189	47	2811
A_2	946	1387	746	53	3132
A_3	115	438	288	16	857
$\nu_{\cdot b}$	2829	2632	1223	116	$\nu=6800$

We have:

$$\nu_{1m} = 1768 \qquad \nu_{m1} = 1768$$
$$\nu_{2m} = 1387 \qquad \nu_{m2} = 1387$$
$$\nu_{3m} = 438 \qquad \nu_{m3} = 746$$
$$\nu_{m4} = 53$$
$$\nu_{\cdot m} = 2829 \qquad \nu_{m\cdot} = 3132$$
$$\lambda_a = \frac{3,954 - 3,132}{6,800 - 3,132} = \frac{822}{3,668} = .2241$$
$$\lambda_b = \frac{3,593 - 2,829}{6,800 - 2,829} = \frac{764}{3,971} = .1924$$
$$\lambda = \frac{822 + 764}{3,668 + 3,971} = \frac{1,586}{7,639} = .2076.$$

(Quotients are given to four places.) The traditional measures of association have the following values: $\chi^2/\nu = .1581$, $C = .3695$, $T = .2541$, Cramér's measure $= .07905$.

This example appears as an illustration of the usual approach to measures of association in [7], a standard statistical reference work. It is not hard to think of interpretations or variations in which one of the λ coefficients would be appropriate. For example, one might be studying the efficacy of an identification scheme for males in which hair color was given but not eye color. Another example might be in connection with a study of popular beliefs about the relationship between hair color and eye color.

5.4. *Weighting Columns or Rows*

In some cases, particularly when comparisons between different populations are important, the measures λ_a, λ_b, or λ may not be suitable, since they depend essentially on the marginal frequencies. To put this in terms of the model of activity: in some cases we do not want to think of choosing an individual from the actual population *at hand* in a random way, but rather from some other population which is related to the actual population in terms of conditional frequencies.

This point is stressed by Yule in reference [15] and is illustrated by the kind of medical example[3] given there. Suppose that we are concerned with the effects of a medical treatment on persons contracting an often fatal disease. Very large samples from two different hospitals are available, giving the following ρ_{ab} tables:

	Hospital I			Hospital II		
	Lived	Died	Total	Lived	Died	Total
Treated	.84	.04	.88	.42	.02	.44
Not treated	.03	.09	.12	.14	.42	.56
Total	.87	.13	1.00	.56	.44	1.00

Here the A classes are Treated or Not-treated, and the B classes Lived or Died. The given numbers are ρ's and marginal ρ's.

We are interested in the association between treatment and life, and might conclude that λ_b would be an appropriate measure of this. We find

$$\lambda_b \text{ for Hospital I} = \frac{.93 - .87}{.13} = .462$$

$$\lambda_b \text{ for Hospital II} = \frac{.84 - .56}{.44} = .636.$$

[3] We do not wish to suggest by this example that λ_b is necessarily appropriate as a measure of association between treatment and cure. A very interesting discussion of this medical case has been given by Greenwood and Yule [3] who bring out many difficulties and suggest various viewpoints. Another interesting paper on the medical 2×2 table is that of Youden [14].

Yet the *conditional* probabilities of life, given treatment (nontreatment), are exactly the same for both hospitals, namely .955 (.250). The reason that the conditional probabilities are the same while the λ_b values are different is, of course, that the two hospitals treated very different proportions of their patients. And the proportions treated were probably determined by factors having nothing to do with 'inherent' association between treatment and cure.

It may seem reasonable in such a case as this to replace our model of activity by one in which an individual is drawn from the population so that the probability of his being in any given A_a is exactly $1/\alpha$, i.e., so that all A classes are equiprobable; and with conditional B class probabilities equal to those of the original population. That is to say, it may seem reasonable to replace the quantities ρ_{ab} by the quantities

$$\frac{1}{\alpha}\frac{\rho_{ab}}{\rho_{a\cdot}} \tag{16}$$

and use this as the population to which λ_b is applied. We may thus define, in terms of the conditional probabilities given A_a,

$$\lambda_b^* = \frac{\dfrac{1}{\alpha}\sum_a \dfrac{\rho_{am}}{\rho_{a\cdot}} - \dfrac{1}{\alpha}\operatorname*{Max}_b \sum_a \dfrac{\rho_{ab}}{\rho_{a\cdot}}}{1 - \dfrac{1}{\alpha}\operatorname*{Max}_b \sum_a \dfrac{\rho_{ab}}{\rho_{a\cdot}}}. \tag{17}$$

If we do this in the present example, we get, of course, the same altered ρ table for both hospitals

.477	.023	.500
.125	.375	.500
.602	.398	1.00

and in both cases

$$\lambda_b^* = \frac{.250}{.398} = .628.$$

An analogous procedure could be used to define λ_a^* and λ^*. Note also that other 'artificial' marginal ρ's besides .5 could be used if appropriate. Yule [15] suggests as a desideratum for coefficients of association their invariance under transformations on the $\{\rho_{ab}\}$ matrix of form

$$\rho_{ab} \rightarrow s_a t_b \rho_{ab}, \; s_a, \; t_b > 0; \quad a = 1, \cdots, \alpha; \quad b = 1, \cdots, \beta.$$

Such a transformation may readily be found (at least when no $\rho_{ab}=0$) to make *all* four marginals of a two by two table equal to .5. In this connection, we refer to a recent article by Pompilj [10] in which such transformations are carefully discussed.

All further measures may be considered for unweighted or weighted marginal proportions, whichever are appropriate.

6. MEASURES BASED UPON OPTIMAL PREDICTION OF ORDER

6.1. *Preliminaries*

Heretofore we have considered measures of association suitable for the unordered case, that is, measures which do not change if the columns (rows) are permuted. Now we shall suggest a measure suitable for the ordered case. Suppose that the situation is of the following kind:

(*i*) Two polytomies, *A* and *B*.
(*ii*) No relevant underlying continua.
(*iii*) Directed ordering is of interest.
(*iv*) The two polytomies appear symmetrically.

By (*iii*) we mean that we wish to distinguish, in the 3×3 case between, for example,

ρ_{11}	0	0
0	ρ_{22}	0
0	0	ρ_{33}

and

0	0	ρ_{13}
0	ρ_{22}	0
ρ_{31}	0	0

calling the first of these complete association and the second complete counterassociation. We may wish to make the convention that in these two cases the proposed measure should take the values $+1$ and -1 respectively. If the sense or direction of order is irrelevant we can, for example, simply take the absolute value of a measure appropriate to directed ordering.

There are vaguenesses in the idea of complete ordered association. For example, everyone would probably agree that the following case is one of complete association:

0	0	0
ρ_{21}	0	0
0	ρ_{32}	0

The following situation is not so clear:

ρ_{11}	0	0
ρ_{21}	ρ_{22}	0
0	ρ_{32}	ρ_{33}

As before, the procedure we shall adopt toward this and toward more complex questions is to base the measure of association on a probabilistic model of activity which often may be appropriate and typical.

6.2. *A Proposed Measure*

Our proposed model will now be described. Suppose that two individuals are taken independently and at random from the population (technically with replacement, but this is unimportant for large populations). Each falls into some (A_a, B_b) cell. Let us say that the first falls in the $(A_{\underline{a}_1}, B_{\underline{b}_1})$ cell, and the second in the $(A_{\underline{a}_2}, B_{\underline{b}_2})$ cell. (Underlined letters denote random variables.) $\underline{a}_i$ $(i=1, 2)$ takes values from 1 to α; $\underline{b}_i$ $(i=1, 2)$ takes values from 1 to β.

If there is independence, one expects that the order of the $\underline{a}$'s has no connection with the order of the $\underline{b}$'s. If there is high association one expects that the order of the $\underline{a}$'s would generally be the same as that of the $\underline{b}$'s. If there is high counterassociation one expects that the orders would generally be different.

Let us therefore ask about the probabilities for like and unlike orders. In order to avoid ambiguity, these probabilities will be taken conditionally on the absence of ties. Set

(18) $\Pi_s = \Pr\{\underline{a}_1 < \underline{a}_2 \text{ and } \underline{b}_1 < \underline{b}_2; \text{ or } \underline{a}_1 > \underline{a}_2 \text{ and } \underline{b}_1 > \underline{b}_2\}$

(19) $\Pi_d = \Pr\{\underline{a}_1 < \underline{a}_2 \text{ and } \underline{b}_1 > \underline{b}_2; \text{ or } \underline{a}_1 > \underline{a}_2 \text{ and } \underline{b}_1 < \underline{b}_2\}$

(20) $\Pi_t = \Pr\{\underline{a}_1 = \underline{a}_2 \text{ or } \underline{b}_1 = \underline{b}_2\}$.

Then the conditional probability of like orders given no ties is $\Pi_s/(1-\Pi_t)$ and the conditional probability of unlike orders given no ties is $\Pi_d/(1-\Pi_t)$. Of course, the sum of these two quantities is one.

A possible measure of association would then be $\Pi_s/(1-\Pi_t)$, but it is a bit more convenient to look at the following quantity:

$$\gamma = \frac{\Pi_s - \Pi_d}{1 - \Pi_t} \tag{21}$$

or the *difference* between the conditional probabilities of like and unlike orders. In other words γ tells us how much more probable it is to get like than unlike orders in the two classifications, when two individuals are chosen at random from the population.

Since $\Pi_s + \Pi_d = 1 - \Pi_t$, we may write γ as

$$\gamma = \frac{2\Pi_s - 1 + \Pi_t}{1 - \Pi_t} \tag{22}$$

which is convenient for computation, using the easily checked relationships

$$\Pi_s = 2 \sum_a \sum_b \rho_{ab} \{ \sum_{a'>a} \sum_{b'>b} \rho_{a'b'} \} \tag{23}$$

$$\Pi_t = \sum_a \rho_{a\cdot}^2 + \sum_b \rho_{\cdot b}^2 - \sum_a \sum_b \rho_{ab}^2. \tag{24}$$

Some important properties of γ follow:

(*i*) γ is indeterminate if the population is entirely in a single row or column of the cross-classification table.

(*ii*) γ is 1 if the population is concentrated in an upper-left to lower-right diagonal of the cross-classification table. γ is -1 if the population is concentrated in a lower-left to upper-right diagonal of the table.

(*iii*) γ is 0 in the case of independence, but the converse need not hold except in the 2×2 case. An example of nonindependence with $\gamma = 0$ is

.2	0	.2
0	.2	0
.2	0	.2

For tables up to 5×5 with ρ's expressed to two decimal places computation is fairly rapid. If many tables of the same size are at hand a cardboard template would be convenient. A check on Π_s is to recompute using inverted ordering in both dimensions. γ may be rewritten in terms of the ν's by putting "ν_{ab}" for "ρ_{ab}," etc., and replacing "1" in (21) and (22) by "ν^2."

In the 2×2 case we find that

$$\gamma = \frac{\rho_{11}\rho_{22} - \rho_{12}\rho_{21}}{\rho_{11}\rho_{22} + \rho_{12}\rho_{21}}. \tag{25}$$

This is the same as Yule's coefficient of association Q mentioned in Section 4. In this case $\gamma = \pm 1$ if any one cell is empty. For example,

ρ_{11}	0
ρ_{21}	ρ_{22}

gives rise to $\gamma = 1$ always.

Any case of the following forms will give rise to $\gamma = 1$, since a conflict in order is impossible:

ρ_{11}	ρ_{12}	0
0	ρ_{22}	ρ_{23}
0	0	ρ_{33}

ρ_{11}	0	0
ρ_{21}	0	0
ρ_{31}	ρ_{32}	ρ_{33}

The right-hand table might be thought of as a case of "complete curvilinear association."

Stuart [11], starting from a suggestion by Kendall [6], has proposed a measure of association in the ordered case much like γ. Stuart's measure, which he calls τ_c is, in our notation,

$$\tau_c = \frac{\Pi_s - \Pi_d}{(m-1)/m}$$

where $m = \text{Min}(\alpha, \beta)$. The term $(m-1)/m$ is introduced in order that τ_c may attain, or nearly attain, the absolute value 1 when the entire population lies in a longest diagonal of the table. Stuart develops his measure by considering a two-way ordered classification table as two rankings of a population, where many ties appear in one or both rankings as two individuals of the population fall in the same column or row or both. Then each ordered pair of individuals is assigned a score with respect to each ranking: 0 if there is a tie, or ± 1 as one or the other is ranked higher. Finally the product-moment correlation coefficient is formally computed with these scores, and the norming factor is introduced.

Thus, our development of γ is seen to give another and more natural interpretation for the numerator of τ_c: it is the probability of like order less the probability of unlike order when two individuals are chosen at random. In addition the form in which τ_c is given above, together with (23) and (24), suggests a computation procedure somewhat different than that of [11].

6.3. *An Example*

Whelpton, Kaiser, and others [17] have investigated in great detail relationships between human fertility and a number of social and psychological characteristics of married couples. The analyses resulting from these investigations are replete with cross-classification tables, together with accompanying verbal explanations and recapitulations. Numerical indexes of association appear to have been used rarely, if at all, in this work.

We wish to examine briefly one of these cross-classification tables as an example of a cross-classification with an order in both classifications. This examination should be construed neither as approval nor criticism of the methodology used in the studies edited by Whelpton and Kaiser, for this would not be appropriate here. (The reader may refer to [18] and [19] for critical reviews.) However, we do feel that the use of summarizing indexes of association in a study of this kind may well be worth while for at least two reasons. One is that the reader finds it very difficult to obtain a bird's-eye view of the extensive numerical material without depending almost wholly on the author's own conclusions. Second, the use of indexes would mitigate the criticism that the author, consciously or not, selects from his numerical data those comparisons that are in line with his a priori beliefs. Needless to say, an index of association is recommended by these arguments only if it has some reasonable interpretation.

The particular table we wish to consider follows, in terms of numbers of married couples. It refers to a rather special, but well defined, population: white Protestant married couples living in Indianapolis, married in 1927, 1928, or 1929, and so on. The data were obtained by stratified sampling, with strata based on numbers of live births. However, for present purposes we do not consider any questions of sampling, response error, specification of population, etc. The table is condensed from a more detailed cross-classification given in [17], vol. 2, pp. 286, 389, and 402. Further, we shall not define the fertility-planning categories that follow, but merely indicate the order.

CROSS-CLASSIFICATION BETWEEN EDUCATIONAL LEVEL OF WIFE AND FERTILITY-PLANNING STATUS OF COUPLE. SOURCE [17], VOL. 2. NUMBERS IN BODY OF TABLE ARE FREQUENCIES

Highest level of formal education of wife	Fertility-planning status of couple				Row totals
	A Most effective planning of number and spacing of children	B	C	D Least effective planning of children	
one year college or more	102	35	68	34	239
3 or 4 years high school	191	80	215	122	608
less than 3 years high school	110	90	168	223	591
Column totals	403	205	451	379	1438

This is clearly a case where there is relevant order in both classifications. We may first compute Π_s as follows (schematically):

$$\Pi_s = \frac{2}{(1438)^2} [102(80 + 90 + 215 + 168 + 122 + 223) + 35(215 + 168 + 122 + 223) + \cdots + 215\,(223)]$$

$$= \frac{2}{(1438)^2} [102 \times 898 + 35 \times 728 + \cdots + 215 \times 223]$$

$$= \frac{2 \times 311{,}632}{2{,}067{,}844} = .301.$$

This means that if we pick two couples at random from those included in the table, the probability is .301 that they are not tied in either classification and that they fall in the same order for both classifications (e.g., if educational level of wife is greater for first couple chosen, then effectiveness of fertility planning is also greater).

Similarly we compute that $\Pi_d = .163$. This is the probability of no ties and *different* orders. Finally Π_t, the probability of a tie in at least one classification, is .536. Note that $\Pi_s + \Pi_d + \Pi_t = 1.000$.

The conditional probability of like order, given no tie, is $\Pi_s/(1 - \Pi_t) = .301/.464 = .649$; and the conditional probability of unlike order is $.163/.464 = .351$. Clearly there is a greater chance of like order than of unlike order, and this means positive association, if the operational model is a reasonable one. To measure the magnitude of this association we may use γ, which here is equal to

$$\frac{.301 - .163}{.464} = .298.$$

This is the difference between the conditional probabilities of like and unlike order, given no ties.

It might be thought that one should look, not at the actual population above, but at a related population with equal row totals and with the same relative frequencies within each row. That is, we might wish to work with a derived population within which one-third of the wives lie in each education category, but which is otherwise the same. This derived population is readily obtained (in terms of its ρ_{ab}'s) by dividing each frequency in the above table by three times the total in its row. Very minor adjustments were made because of rounding, in order that the over-all sum be 1.000. For the same reason, the row totals are not exactly equal.

CROSS-CLASSIFICATION BETWEEN EDUCATIONAL LEVEL OF WIFE AND FERTILITY-PLANNING STATUS OF COUPLE. DERIVED FROM PRIOR TABLE BY ADJUSTMENT TO MAKE ROW TOTALS EQUAL. NUMBERS IN BODY OF TABLE ARE RELATIVE FREQUENCIES (ρ_{ab}'s).

Highest level of formal education of wife	Fertility-planning status of couple				
	A Most effective planning of number and spacing of children	B	C	D Least effective planning of children	Row totals
one year college or more	.142	.049	.095	.047	.333
3 or 4 years high school	.105	.044	.118	.067	.334
less than 3 years high school	.062	.050	.095	.126	.333
Column totals	.309	.143	.308	.240	1.000

For this table we find $\Pi_s = .325$, $\Pi_d = .170$, $\Pi_t = .505$.

Hence $\Pi_s/(1-\Pi_t) = .657$, $\Pi_d/(1-\Pi_t) = .343$, and $\gamma = .314$. There is no great difference between the original and the adjusted table in regard to association as measured by probabilities of like and unlike order.

Alternatively, one might wish to adjust the tabular entries so that column totals are equal, or one might attempt to adjust the entries so that the row totals are equal and the column entries are equal.

7. THE GENERATION OF MEASURES BY THE INTRODUCTION OF LOSS FUNCTIONS

7.1. *Models Based on Loss Functions*

Instead of obtaining a measure as a natural function of probabilities in the context of a model of predictive behavior, one can more generally employ loss functions. In such a way, one can even artificially generate the conventional measures described in Section 4.

7.2. *Loss Functions and the λ Measures*

In the context of Section 5.1 let us suppose that in guessing an individual's B class one incurs a loss $L(b_1, b_2)$, where B_{b_1} is the true B class and B_{b_2} is the guessed one. Consider first guessing B_b given no information. Then a scheme of guessing B_b with probability $p_b(p_b \geqq 0, \sum p_b = 1)$ leads to an average loss of $\sum_{b_1} \sum_{b_2} \rho_{\cdot b_1} p_{b_2} L(b_1, b_2)$. It is easily seen that this average is minimized by guessing that B_{b_2} for which $\sum_b \rho_{\cdot b} L(b, b_2)$ is a minimum, or if there are two or more minima by guessing any one of them. Let b_L be any one of these b_2's, so that the minimum average loss is $\sum_b \rho_{\cdot b} L(b, b_L)$.

On the other hand if the individual's A class is known to be A_a, the best scheme of guessing is to select b_2 to minimize $\sum_b \rho_{ab} L(b, b_2)$. Let b_{La} be such a minimizing b_2; then the minimum average loss when A_a is known is $\sum_b (\rho_{ab}/\rho_{a\cdot}) L(b, b_{La})$, and the over-all minimum average loss with A_a's known is $\sum_a \sum_b \rho_{ab} L(b, b_{La})$.

The decrease in loss as we pass from the first case to the second is therefore

$$\sum_b \rho_{\cdot b} L(b, b_L) - \sum_a \sum_b \rho_{ab} L(b, b_{La}). \tag{26}$$

It would be reasonable to norm this by division by the first term, $\sum_b \rho_{\cdot b} L(b, b_L)$, to obtain a generalization of λ_b.

Notice that if $L(b_1, b_2)$ is 0 when $b_1 = b_2$ and 1 when $b_1 \neq b_2$, we obtain exactly λ_b. Analogous procedures give us generalizations of λ_a and λ. A slight extension of the procedure, permitting the loss to depend on the true A class as well as the true and guessed B classes, gives a generalization of λ_b^*.

7.3. *The Conventional Measures in Terms of Loss Functions*

Suppose, instead of predicting the classes of individuals, we are asked to determine the values ρ_{ab} when only the $\rho_{a\cdot}$ and $\rho_{\cdot b}$ are known. In the case of independence, these ρ_{ab} are $\rho_{a\cdot}\ \rho_{\cdot b}$. In the more general case, the difference between ρ_{ab} and $\rho_{a\cdot}\ \rho_{\cdot b}$ may be thought of as the amount of error made by assuming independence, If the loss is proportional to the square of the error, inversely proportional to the estimate $\rho_{a\cdot}\ \rho_{\cdot b}$, and additive, we have

$$\sum_a \sum_b k_{ab} \frac{(\rho_{ab} - \rho_{a\cdot}\rho_{\cdot b})^2}{\rho_{a\cdot}\rho_{\cdot b}} \tag{27}$$

where the k_{ab}'s are given constants. For comparison with standard chi-square, express this in terms of the ν_{ab}'s

$$\sum_a \sum_b k_{ab} \frac{\left(\nu_{ab} - \frac{\nu_{a\cdot}\nu_{\cdot b}}{\nu}\right)^2}{\nu_{a\cdot}\nu_{\cdot b}} \tag{28}$$

and finally set $k_{ab} = \nu$ to obtain just the chi-square statistic.

Although this procedure and loss function seem to us rather artificial, they do give one way of motivating the chi-square statistic as a measure of association.

8. RELIABILITY MODELS

8.1. *Generalities*

Consider now cases in which the classes are the *same* for the two polytomies, so that we deal with an $\alpha \times \alpha$ table, but differ in that assignment to class depends on which of two methods of assignment is used. Thus we might for example consider two psychological tests both of which classify deranged individuals as to the type of mental disorder from which they suffer. Or again, we might consider two observers taking part in a sociological experiment wherein they independently and subjectively rate each child in a group of children on a five point scale for degree of cooperation.

One is often concerned in such cases with the degree to which the two methods of assignment to class agree with each other. In the case of the psychological tests, for example, one of the tests might be a well established standard procedure and the other might be a more easily applied variant under consideration as a substitute. The psychologist would probably only consider the variant seriously if it gave the same answers as the standard test often enough in some sense which he would have to explicate. In the case of the two observers, the problem might be whether the kind of subjective ratings given by trained observers in that context are similar enough to warrant the use of such subjective ratings at all.

As before we shall not consider here sampling problems, but rather shall suppose the population ρ_{ab}'s known. The several distinctions and conventions of Sections 2 and 3 apply here of course, but the measures suggested in Sections 5 and 6 do not seem appropriate in this reliability context. One reason is that the classes are the same for both polytomies. This means that even in the unordered case we do *not* want a measure which is invariant under interchange of rows and interchange of columns unless the two interchanges are the same.

An obvious measure of reliability in such a study is just $\sum_a \rho_{aa}$, the probability of agreement. However, we shall also consider some other possibilities.

8.2. *A Measure of Reliability in the Unordered Case*

The measure we shall now propose might be appropriate under the following conditions:

(*i*) Two polytomies are the same, but arise from different methods of assignment to class.
(*ii*) No relevant underlying continua.
(*iii*) No relevant ordering.
(*iv*) Our interest in reliability is symmetrical as between the two polytomies.

A modal class over both classifications is any $A_a(=B_a)$ such that $\rho_{a\cdot}+\rho_{\cdot a} \geqq \rho_{a'\cdot}+\rho_{\cdot a'}$ for all a'. It is simplest to suppose that there is a unique modal class, but if there are more any can be chosen. Denote by $\rho_{M\cdot}$ and $\rho_{\cdot M}$ the two marginal proportions corresponding to the modal class.

A modal class can be given the following interpretation: choose an individual at random from the population and pick one of the two methods of assignment by flipping a fair coin. What is the long-run best guess beforehand of how the chosen method will classify the chosen individual? The answer is: the modal class; and if the modal class is A_a, then the probability of a correct guess is $\frac{1}{2}(\rho_{a\cdot}+\rho_{\cdot a})=\frac{1}{2}(\rho_{M\cdot}+\rho_{\cdot M})$.

In so far as there is good reliability between the two methods of assignment, one could make a better guess if one knew how the other method of assignment would classify the individual, and then followed the rule of guessing the *same* class for the method being predicted. The probability of a correct guess would then be $\sum \rho_{aa}$. Thus as we go from the no information situation to the other-method-known situation, the probability of error decreases by $\sum \rho_{aa}-\frac{1}{2}(\rho_{M\cdot}+\rho_{\cdot M})$. This quantity may vary from $-\frac{1}{2}$ to $1-(1/\alpha)$. It takes the value $-\frac{1}{2}$ when all the diagonal ρ_{aa}'s are zero and the modal probability, $\rho_{M\cdot}+\rho_{\cdot M}$ is 1. It takes the value $1-(1/\alpha)$ when the two methods always agree and each category is equi-probable.

To get a measure we should alter the above quantity, since a sufficiently large ρ_{aa} for some a will make the above quantity low even though $\sum \rho_{aa}$ is nearly 1. It seems reasonable to norm by division by the probability of error given no information, that is by $1-\frac{1}{2}(\rho_{M\cdot}+\rho_{\cdot M})$. Hence we propose the measure

$$(29) \qquad \lambda_r = \frac{\sum \rho_{aa} - \frac{1}{2}(\rho_{M\cdot} + \rho_{\cdot M})}{1 - \frac{1}{2}(\rho_{M\cdot} + \rho_{\cdot M})}.$$

This may be interpreted as the relative decrease in error probability as we go from the no information situation to the other-method-known situation.

* The measure λ_r can take values from -1 to 1. It takes the value -1 when all the diagonal ρ_{aa}'s are zero and the modal probability, $\rho_{M\cdot}+\rho_{\cdot M}$ is 1. It takes the value 1 when the two methods always agree. λ_r is indeterminate only when both methods always give only one and the same class. In the case of independence λ_r assumes no particular value. This characteristic might be considered a disadvantage, but it seems to us that an index of this kind would only be used where there is known to be dependence between the methods, so that misbehavior of the index for independence is not important.

8.3. *Reliability in the Ordered Case*

For the case in which the classes are ordered, but a meaningful metric is absent, we have been unable to find a measure better than one of the following kind:

$$(30a) \qquad \sum_{a=1}^{\alpha} \rho_{aa} \quad \text{(as suggested in Section 8.1)}$$

$$(30b) \qquad \sum_{|a-b|\leq 1} \rho_{ab}$$

$$(30c) \qquad \sum_{|a-b|\leq 2} \rho_{ab},$$

that is, the only reasonable measures we know of are those that are based upon either the probability of agreement, the probability of agreement to within one neighboring class, two neighboring classes, and so on. If desired one could weight these probabilities when classification in a neighboring class is not as desirable as in the same class. Thus one might consider something like $\sum \rho_{aa} + \frac{1}{2} \sum_{|a-b|=1} \rho_{ab}$ or its obvious variants. These measures may also be justified easily by loss-function arguments.

9. PROPORTIONAL PREDICTION

Instead of basing a measure of association on optimal prediction one might consider measures based upon a prediction method which reconstructs the population, in a sense to be described. The use of such a measure was suggested to us by W. Allen Wallis. For simplicity, we restrict ourselves to the asymmetric situation of Section 5.1 where λ_b was constructed. Of course one could apply the same approach in other situations.

Our model of activity, as before, is the following: An individual is chosen at random from the population and we are asked to guess his B class either (1) given no information or (2) given his A class.

Optimal guessing will lead to a definite B class in case (1) and to a definite B class for each A class in case (2) (except that in the case of tied $\rho_{\cdot b}$'s or ρ_{ab}'s we have some choice). While such optimal guessing leads to the lowest average frequency of error, the resulting distribution of guessed classes will usually be very different from the original distribution in the population. For some purposes this might be undesirable and one is led to the following model of activity:

Case *1.* Guess B_1 with probability $\rho_{\cdot 1}$, B_2 with probability $\rho_{\cdot 2}$, $\cdots$, B_β with probability $\rho_{\cdot \beta}$.

Case *2.* Guess B_1 with probability $\rho_{a1}/\rho_{a\cdot}$ (the conditional probability of B_1 given A_a), B_2 with probability $\rho_{a2}/\rho_{a\cdot}$, $\cdots$, B_β with probability $\rho_{a\beta}/\rho_{a\cdot}$.

In each case the guessing is to proceed by throwing a β-sided die whose bth side appears with probability $\rho_{\cdot b}$ (case 1) or $\rho_{ab}/\rho_{a\cdot}$ (case 2). This may be accomplished using a table of "random numbers." If we make many such guesses independently it is plain that we shall approximately reconstruct the marginal distribution of the B_b's (case 1) and the joint distribution of the (A_a, B_b)'s (case 2).

The long-run proportion of *correct* predictions in case (1) will be $\sum_{b=1}^{\beta} \rho_{\cdot b}^2$, and in case (2) it will be $\sum_{a=1}^{\alpha} \sum_{b=1}^{\beta} \rho_{ab}^2/\rho_{a\cdot}$. Hence the relative *decrease* in the proportion of incorrect predictions as we go from case (1) to case (2) is

$$\tau_b = \frac{\sum_a \sum_b \rho_{ab}^2/\rho_{a\cdot} - \sum_b \rho_{\cdot b}^2}{1 - \sum_b \rho_{\cdot b}^2} \tag{31}$$

which can be readily expressed in the chi-square-like form

$$\tau_b = \frac{\sum_a \sum_b \frac{(\rho_{ab} - \rho_{a\cdot}\rho_{\cdot b})^2}{\rho_{a\cdot}}}{1 - \sum_b \rho_{\cdot b}^2}, \tag{32}$$

It is clear that τ_b takes values between 0 and 1; it is 0 if and only if there is independence, and 1 if and only if knowledge of A_a completely determines B_b. Finally τ_b is indeterminate if and only if both independence and determinism simultaneously hold, that is if all $\rho_{\cdot b}$'s but one are zero.

10. ASSOCIATION WITH A PARTICULAR CATEGORY

A group of modifications of many of the preceding measures arises from the observation that there may be little association between the A and B polytomies in general, but if an individual is in a particular A class it may be easy to predict his B class. Suppose, then, that we want the association between A_{a_0}, a specific A class, and the B polytomy. One need only condense all the A_a rows where $a \neq a_0$ into a single row, thus obtaining a $2 \times \beta$ table, and apply whatever measure of association is thought appropriate. The table will have this appearance.

	B_1	B_2	$\cdots$	B_β
A_{a_0}	$\rho_{a_0 1}$	$\rho_{a_0 2}$	$\cdots$	$\rho_{a_0 \beta}$
$A_a\ (a \neq a_0)$	$\rho_{\cdot 1} - \rho_{a_0 1}$	$\rho_{\cdot 2} - \rho_{a_0 2}$	$\cdots$	$\rho_{\cdot \beta} - \rho_{a_0 \beta}$

We are indebted to L. L. Thurstone for pointing out to us the importance of this modification.

11. PARTIAL ASSOCIATION

When there are more than two polytomies it is natural to think of partial association between two of them with the effect of the others averaged out in some sense. Two such measures of partial association will be suggested here for the asymmetrical case and three polytomies. The viewpoint will be that of optimal prediction. Analogous symmetrical measures may be readily obtained, and the restriction to three polytomies is purely for convenience of notation. The first two polytomies will be denoted as before; the third will consist of the classification $C_1, C_2, \cdots, C_\gamma$. The proportion of the population in A_a, B_b, and

C_c is ρ_{abc}, and dots will be used to denote marginal values in the conventional way. The proposed measures will be for partial association between the A and B polytomies 'averaged' over the C polytomy. (Do not confuse the integer γ used here with the index γ of Section 6.)

11.1. *Simple Average of* λ_b

For fixed C_c, we have a conditional $A \times B$ double polytomy with relative frequencies $\rho_{abc}/\rho_{\cdot\cdot c}$. Hence we can compute λ_b for each such table—call it $\lambda_b(c)$ to show dependence on c. Now it might seem natural to average these values with weights equal to the marginal relative frequencies of the C classifications. That is, we suggest

$$\text{(33)} \qquad \lambda_b(A, B \mid C) = \sum_{c=1}^{\gamma} \rho_{\cdot\cdot c}\lambda_b(c).$$

11.2. *Measure Based Directly on Probabilities of Error*

It seems to us somewhat better, from the viewpoint of interpretation, to proceed as follows. For given C_c if we predict B classes optimally on the basis of no further information, the probability of error is $1-(\text{Max}_b\, \rho_{\cdot bc})/\rho_{\cdot\cdot c}$; whereas if we know the A class the probability of error is $1-(\sum_a \text{Max}_b\, \rho_{abc})/\rho_{\cdot\cdot c}$. Hence, if we are given individuals from the population at random and always told their C class, the probability of error in optimal guessing if we know nothing more is $1-\sum_c \text{Max}_b\, \rho_{\cdot bc}$; whereas if we also know the A class the probability is $1-\sum_c \sum_a \text{Max}_b\, \rho_{abc}$. Thus the relative decrease in probability of error is

$$\text{(34)} \qquad \lambda_b'(A, B \mid C) = \frac{\sum_c \sum_a \text{Max}_b\, \rho_{abc} - \sum_c \text{Max}_b\, \rho_{\cdot bc}}{1 - \sum_c \text{Max}_b\, \rho_{\cdot bc}}$$

which might often be a satisfactory measure of partial association.

12. MULTIPLE ASSOCIATION

When there are more than two polytomies one may well be interested in the multiple association between one of them and all the others. One simple way of handling this in the unordered case will be described here for three polytomies A, B, and C as defined in Section 11. We suppose that the multiple association between A and B-together-with-C is of interest. Simply form a two-way table whose rows represent the A polytomy and whose columns represent all combinations B_b, C_c and then apply the appropriate two-polytomy measure. The table will have this appearance:

	B_1C_1	B_1C_2	$\cdots$	B_1C_γ	B_2C_1	$\cdots$	B_2C_γ	$\cdots$	$B_\beta C_\gamma$
A_1	ρ_{111}	ρ_{112}	$\cdots$	$\rho_{11\gamma}$	ρ_{121}	$\cdots$	$\rho_{12\gamma}$	$\cdots$	$\rho_{1\beta\gamma}$
A_2	ρ_{211}	ρ_{212}	$\cdots$	$\rho_{21\gamma}$	ρ_{221}	$\cdots$	$\rho_{22\gamma}$	$\cdots$	$\rho_{2\beta\gamma}$
$\vdots$	$\vdots$	$\vdots$	$\cdots$	$\vdots$	$\vdots$	$\cdots$	$\vdots$	$\cdots$	$\vdots$
A_α	$\rho_{\alpha 11}$	$\rho_{\alpha 12}$	$\cdots$	$\rho_{\alpha 1\gamma}$	$\rho_{\alpha 21}$	$\cdots$	$\rho_{\alpha 2\gamma}$	$\cdots$	$\rho_{\alpha\beta\gamma}$

Note that this procedure does not take the $B\times C$ association into account. There is a rough analogy here with the motivation for the standard multiple correlation coefficient of normal theory. The standard multiple correlation coefficient may be (and often is) motivated by defining it as the maximum correlation coefficient obtainable between a given variate and linear combinations of the other variates. That is, it is a measure of association between a given variate and the best estimate (in a certain sense) of that variate based upon all the other variates. It is true that the standard multiple correlation coefficient may be expressed as a function of the several ordinary bivariate correlation coefficients, but in a sense this is a consequence of the strong structural assumption of multivariate normality.

13. SAMPLING PROBLEMS

The discussion thus far has been in terms of *known* populations, whereas in practice one generally deals with a sample from an *unknown* population. One then asks, given a formal measure of association, how to estimate its value, how to test hypotheses about it, and so on.

Exact sampling theory for estimators from cross-classification tables is difficult to work with. However, the asymptotic theory is reasonably manageable, at least in some cases. We intend to discuss this in another paper, where we shall state some of the asymptotic distributions and say what we can of their value as approximations.

14. CONCLUDING REMARKS

The aim of this paper has been to argue that measures of association should not be taken blindly from the handiest statistics textbook, but rather should be carefully constructed in a manner appropriate to the problem at hand. To emphasize and illustrate this argument we have described a number of such measures which we feel might be useful in several situations. While we naturally take a friendly view towards these measures, we can hardly claim that they are more than examples.

This methodologically neutral position should not be carried to an extreme. It would be ridiculous to ask each empirical scientist in each separate study to forge afresh new statistical tools. The artist cannot paint many pictures if he must spend most of his time mixing pigments. Our belief is that each scientific area that has use for measures of association should, after appropriate argument and trial,[4] settle down on those measures most useful for its needs.

15. REFERENCES

[1] Cramér, Harald, *Mathematical Methods of Statistics*, Princeton, Princeton University Press (1946)

[2] Fisher, R. A., *Statistical Methods for Research Workers*, Tenth Edition, New York, Hafner Publishing Co. (1948)

[3] Greenwood, Major, Jr., and Yule, G. Udny, "The statistics of anti-typhoid and anti-cholera inoculations, and the interpretation of such statistics in general," *Proceedings of the Royal Society of Medicine*, 8 [part 2] (1915), 113–94.

[4] Guttman, Louis, "An outline of the statistical theory of prediction," Supplementary Study B-1 (pp. 253–318) in Horst, Paul and others (editors), *The Prediction of Personal Adjustment*, Bulletin 48, Social Science Research Council, New York (1941).

[5] Jahn, Julius A., "The measurement of ecological segregation: derivation of an index based on the criterion of reproducibility," *American Sociological Review*, 15 (1950), 101–4.

[6] Kendall, M. G., "Rank and product-moment correlation," *Biometrika*, 36 (1949), 177–93.

[7] Kendall, Maurice G., *The Advanced Theory of Statistics*, London, Charles Griffin and Co., Ltd. (1948).

[8] McCormick, Thomas C., "Toward causal analysis in the prediction of attributes," *American Sociological Review*, 17 (1952), 35–44.

[9] Pearson, Karl, and Heron, David, "On theories of association," *Biometrika*, 9 (1913), 159–315.

[10] Pompilj, G., "Osservazioni sull'omogamia: La trasformazione di Yule e il limite della trasformazione ricorrente di Gini," *Rendiconti di Matematica e*

[4] For examples of such argument and trial in the field of sociology see J. J. Williams [13], Jahn [5], and McCormick [8].

delle sue Applicazioni, Università di Roma, Istituto Nazionale di Alta Matematica, Ser. V, Vol. 9 (1950), 367–88.

[11] Stuart, A., "The estimation and comparison of strengths of association in contingency tables," *Biometrika,* 40 (1953), 105–10.

[12] Williams, E. J., "Use of scores for the analysis of information in contingency tables," *Biometrika,* 39 (1952), 274–89.

[13] Williams, Josephine J., "Another commentary on so-called segregation indices," *American Sociological Review,* 13 (1948), 298–303.

[14] Youden, W. J., "Index for rating diagnostic tests," *Cancer,* 3 (1950), 32–5.

[15] Yule, G. Udny, "On the methods of measuring association between two attributes," *Journal of the Royal Statistical Society,* 75 (1912), 579–642.

[16] Yule, G. Udny, and Kendall, M. G., *An Introduction to the Theory of Statistics,* London, Charles Griffin and Co., Ltd. (1950)

[17] Whelpton, P. K., and Kiser, Clyde V., *Social and Psychological Factors Affecting Fertility,* Milbank Memorial Fund, New York.

Volume 1 (1946), *The Household Survey in Indianapolis.*

Volume 2 (1950), *The Intensive Study; Purpose, Scope, Methods, and Partial Results.*

Volume 3 (1952), *Further Reports on Hypotheses in the Indianapolis Study.*

[18] Jaffe, A. J., review of [17], Volume 2, *Journal of the American Statistical Association,* 47 (1952), 348–9.

[19] Lewis-Faning, E., review of [17], Volume 3, *Journal of the American Statistical Association,* 49 (1954), 190–3.

Reprinted from the JOURNAL OF THE AMERICAN STATISTICAL ASSOCIATION
March 1959, Vol. 54, pp. 123-163

MEASURES OF ASSOCIATION FOR CROSS CLASSIFICATIONS. II: FURTHER DISCUSSION AND REFERENCES*

LEO A. GOODMAN AND WILLIAM H. KRUSKAL
University of Chicago

1. INTRODUCTION AND SUMMARY 124
2. SUPPLEMENTARY DISCUSSION TO PRIOR PAPER 125
2.1. Cross classifications in which the diagonal is not of interest 125
2.2. A relation between the λ measures and Yule's Y 125
2.3. Symmetrical variant of proportional prediction 125
2.4. Association with a particular set of categories 126
2.5. Comparison of degrees of association exhibited by two cross classifications 126
2.6. A new measure of association in the latent structure context 127
2.7. Two corrections 127
3. WORK ON MEASURES OF ASSOCIATION IN THE LATE NINETEENTH AND EARLY TWENTIETH CENTURIES 127
3.1. Doolittle, Peirce, and contemporary Americans; Köppen 127
3.2. Körösy, Jordan, and Quetelet 132
3.3. Benini 133
3.4. Lipps 135
3.5. Tönnies 135
3.6. Deuchler 135
3.7. Gini 136
4. MORE RECENT PUBLICATIONS 137
4.1. Textbook discussions. Guilford, Dornbusch and Schmid, Wallis and Roberts 137
4.2. Reliability measures. Wood, Reuning, Cartwright 138
4.3. Measures that are zero if and only if there is independence. Cramér, Steffensen, Pollaczek-Geiringer, Höffding, Eyraud, Fréchet, Féron 139
4.4. Measures of dissimilarity, especially in the $\alpha \times 2$ case. Gini, Florence, Hoover, Duncan and Duncan, Bogue, Boas, Long, Loevinger 143
4.5. Measures based on Lorenz or cost-utility curves 146
4.6. Measures based on Shannon-Wiener information. McGill, Holloway, Woodbury, Wahl, Linfoot, Halphen 147
4.7. Recent proposals by Italian authors other than Gini. Salvemini, Bonferroni, Brambilla, Faleschini, Andreoli 147

---

* This research was supported in part by the Army, Navy, and Air Force through the Joint Services Advisory Committee for Research Groups in Applied Mathematics and Statistics, Contract No. N6ori-02035; and in part by the Office of Naval Research. This paper, in whole or in part, may be reproduced for any purpose of the United States Government. Part of Mr. Kruskal's work was done at the Department of Statistics, University of California, Berkeley.

We wish to thank the following persons for their generous suggestions and criticisms: J. Berkson (Mayo Clinic), C.E. Bonferroni (University of Florence), G. W. Brier (U. S. Weather Bureau), H. Byers (University of Chicago), F. R. Dodge (Berkeley, Calif.), H. E. Driver (Indiana University), O. D. Duncan (University of Chicago), C. Eisenhart (National Bureau of Standards), L. Faleschini (University of Milan), R. Féron (University of Paris), M. Fréchet (University of Paris), C. Gini (University of Rome), I. I. Gringorten (Air Force Cambridge Research Center), L. Guttman (Israel Institute of Applied Research), C. Jordan (University of Budapest), A. L. Kroeber (University of California), P. F. Lazarsfeld (Columbia University), H. V. Roberts (University of Chicago), T. Salvemini (University of Rome), I. R. Savage (University of Minnesota), R. H. Somers (Columbia University), J. W. Tukey (Princeton University), M. Uridge (University of California), W. van der Bijl (Kansas State College), E. W. Wahl (Air Force Cambridge Research Center), H. Walker (Teachers College, Columbia University), W. A. Wallis (University of Chicago), M. A. Woodbury (New York University).

4.8. Problems of inference discussed by Wilson, Berkson, and Mainland........ 148
4.9. Measures based on latent structures. Lazarsfeld and Kendall............ 148
4.10. More recent work on measures of association in meteorology. Gringorten, Bleeker, Brier, and others... 152
4.11. Association between species. Forbes, Cole, Goodall.................... 153
4.12. Association between anthropological traits. Tylor, Clements, Wallis, Driver, Kroeber, Chrétien, Kluckhohn, and others...................... 154
4.13. Other suggestions. Harris, Pearson, Irwin, Lakshmaramurti, Fairfield Smith 155
REFERENCES... 156

Our earlier discussion of measures of association for cross classifications [66] is extended in two ways. First, a number of supplementary remarks to [66] are made, including the presentation of some new measures. Second, historical and bibliographical material beyond that in [66] is critically surveyed; this includes discussion of early work in America by Doolittle and Peirce, early work in Europe by Körösy, Benini, Lipps, Deuchler and Gini, more recent work based on Shannon-Wiener information, association measures based on latent structure, and relevant material in the literatures of meteorology, ecology, sociology, and anthropology. New expressions are given for some of the earlier measures of association.

1. INTRODUCTION AND SUMMARY

THIS paper has two purposes. First, we wish to present a supplementary discussion to problems considered in our first paper on cross classifications [66], including presentation of some new measures; this is Section 2 of the present paper. Second, we wish to extend the brief historical and bibliographical remarks in [66] to include a number of publications, many of them little-known, that may be of interest to those working with cross classifications; this is done in Sections 3 and 4 of the present paper.

We have in preparation a paper on approximate distributions for the sample analogues of the measures of association described in [66], but it seems desirable to bring the present remarks, virtually none of which deal with sampling distributions, to the reader's attention in a separate report.

The literature on measures of association for cross classifications is vast, it is poorly integrated, and seldom in this literature are meaningful interpretations of measures adduced. One finds the same questions discussed in papers on meteorology, anthropology, ecology, sociology, etc. with hardly any cross references and with considerable duplication. In surveying this literature, we have been selective, although the length of this paper may not suggest it. Discussion of a measure of association here does not mean *ipso facto* that it has an operational interpretation, a very desirable characteristic for which we argued in [66], but may simply reflect some other interesting aspect of the measure, for example its historical role.

One may organize the historical and bibliographical material in various ways, classifying by date, by type of measure, by substantive field, and so on. We have used a gross chronological division, but within it we have classified in several ways, as seemed most appropriate. Material from [66] has not been repeated here.

2. SUPPLEMENTARY DISCUSSION TO PRIOR PAPER

2.1. *Cross classifications in which the diagonal is not of interest.* Herbert Goldhamer (Rand Corporation) has been concerned with measuring association for $\alpha \times \alpha$ tables where the classes are the same for the two polytomies, as in Section 8 of [66], but where the diagonal entries are of little or no interest. For example, one might tabulate occupation of father against occupation of son, and investigate the association between the two occupations only in the off-diagonal subpopulation where they are not the same. Thus the situation, while similar to that of reliability measures, as in Section 8 of [66], differs from it in that the diagonal entries must not play a part; and hence λ_r of [66] would not be suitable.

It seems to us that reasonable measures of association in this situation would be provided by λ_a, λ_b, or λ in the unordered case, and by γ in the ordered case, when these measures are applied to the conditional classification with all $\rho_{aa}=0$. Hence, replacing ρ_{ab}, for $a \neq b$, by $\rho_{ab}/(1-\sum \rho_{aa})$, and taking all $\rho_{aa}=0$, we would get a new table for which the λ's or γ would have direct conditional interpretations. This kind of simple modification is often easy to make for measures with operational interpretations, whereas it is not at all clear how one might usefully alter a chi-square-like measure to fit Goldhamer's problem. A similar point is made in another context in Section 4.13.

2.2. *A relation between the λ measures and Yule's Y.* Suppose that in the 2×2 case we make a transformation of form $\rho_{ab} \to s_a t_b \rho_{ab}$ so that all the marginals become .5 [66, Sec. 5]. Then, for the altered table, $\lambda_a = \lambda_b = \lambda$, and all three are equal to the absolute value of Y, where

$$Y = \frac{\sqrt{\rho_{11}\rho_{22}} - \sqrt{\rho_{12}\rho_{21}}}{\sqrt{\rho_{11}\rho_{22}} + \sqrt{\rho_{12}\rho_{21}}} \qquad (\rho\text{'s of original table})$$

as described in Section 4 of [66]. The actual transformation is that for which

$$(s_1 : s_2 : t_1 : t_2) = (\sqrt{\rho_{21}\rho_{22}} : \sqrt{\rho_{11}\rho_{12}} : \sqrt{\rho_{12}\rho_{22}} : \sqrt{\rho_{11}\rho_{21}}).$$

Thus we have another formal identity in the 2×2 case between a classical measure of association and one with an operational interpretation.

2.3. *Symmetrical variant of proportional prediction.* In Section 9 of [66], we mentioned a measure of association based, not on optimal prediction, but on proportional prediction in a manner there explained. If one predicts polytomy B half the time and polytomy A the other half, always using proportional prediction, then the relative decrease in the proportion of incorrect predictions, as one goes from the nothing-given situation to the other-polytomy-category-given situation, is

$$\frac{\frac{1}{2}\sum_a \sum_b \{(\rho_{ab} - \rho_{a\cdot}\rho_{\cdot b})^2(\rho_{a\cdot} + \rho_{\cdot b})/(\rho_{a\cdot}\rho_{\cdot b})\}}{1 - \frac{1}{2}\sum_a \rho_{a\cdot}^2 - \frac{1}{2}\sum_b \rho_{\cdot b}^2}.$$

In the 2×2 case this quantity, together with the asymmetrical τ_b of [66], reduces to

$$\frac{(\rho_{11}\rho_{22} - \rho_{12}\rho_{21})^2}{\rho_{1\cdot}\rho_{2\cdot}\rho_{\cdot 1}\rho_{\cdot 2}},$$

or ϕ^2, the mean square contingency.

2.4. *Association with a particular set of categories.* In Section 10 of [66], we described a simple way to consider association between a *particular* A category and the B polytomy; namely coalescence of the $\alpha \times \beta$ table into a $2 \times \beta$ table whose rows correspond to the particular A category and its negation respectively. A similar suggestion was made by Karl Pearson in 1906 [112].

We now discuss association between a particular set of A categories and the B polytomy. Suppose that we want to consider the association between $A_{a_1}, A_{a_2}, \cdots, A_{a_s}$, a specific set of A classes, and the B polytomy. One possible approach is to condense all the A_a rows that are not in the specific set of A classes (*i.e.*, all the A_a rows where a is not equal to any a_k, $k=1, 2, \cdots, s$) into a single row, thus obtaining an $(s+1)\times\beta$ table, and then apply whatever measure of association is thought appropriate. This approach might be used if the entire original population is of interest, and we are only concerned with association for the specific set of A categories and their pooled remainder. If, however, the population of interest consists *only* of those individuals who are in the specific set of A categories, $A_{a_1}, A_{a_2}, \cdots, A_{a_s}$, $(s \geq 2)$, then we would apply whatever measures of association are thought appropriate (e.g., λ_a, λ_b, λ, γ, etc.) to the conditional classification with $\rho_{ab}=0$ for all a that are not equal to any a_k, $k=0, 1, \cdots, s$. That is, we would delete all rows except those corresponding to $A_{a_1}, \cdots, A_{a_s}$, and in those rows we would replace ρ_{ab} by $\rho_{ab}/\sum_{k=1}^{s}\sum_{b=1}^{\beta}\rho_{a_k b}$. We would then have an $s\times\beta$ table, and the λ's or γ would have direct conditional interpretations.

The association between a particular set of A categories and a particular set of B categories, or a particular set of combined (grouped) A categories and a particular set of combined B categories, can be treated in an analogous manner.

2.5. *Comparison of degrees of association exhibited by two cross classifications.* Sometimes one wishes to compare the degrees of association shown by two cross-classified populations. This question is particularly likely to arise when the two classifications are the same for both populations. It was discussed briefly on page 740 of [66]; a bit more detail may be of interest here.

Suppose, for example, that we are considering two populations, each cross classified by the same pair of polytomies and such that λ_b is the appropriate measure of association. That is, the relative decrease in probability of error for optimum prediction of column, as we go from the case of row unknown to that of row known, is the relevant population characteristic. Then the difference between the λ_b's of the two populations gives a simple comparison with a clear meaning. Sometimes the *relative* difference between the λ_b's might be of more interest.

If the pairs of classifications for the two populations are not identical, as will necessarily be the case when the two cross classification tables are of different sizes, the purpose of comparison may not be clear. However, the absolute or relative differences described above may still be used and have perfectly definite interpretations. Of course, the above comments are applicable, not only to λ_b, but to any other measure of association that has an operational meaning.

When we are concerned with sampling problems, the question may arise whether two sample values of λ_b (say) from two different populations differ with statistical significance. This question, together with other questions relating to sampling, will be considered in a paper now in preparation. In that paper we shall also discuss the question of whether K sample values of λ_b from K different populations ($K \geq 2$) differ with statistical significance.

2.6. *A new measure of association in the latent structure context.* Several measures of association discussed in Sections 3 and 4 are based upon probabilistic models of a latent structure nature. This kind of model is explained and discussed in Section 4.9, and there we suggest a new measure in addition to those already suggested by others.

2.7. *Two corrections.* The second and third sentences of the second paragraph of [66], p. 758, are essentially correct, but may be misleading. It would have been clearer to have written

> It [λ_r] takes the value -1 if and only if (i) all ρ_{ab}'s not in the row or column of the modal class are zero, and (ii) ρ_{aa} for the modal class is not one. It takes the value 1 if and only if (i) $\Sigma\rho_{aa} = 1$ (i.e. the two methods always agree), and (ii) ρ_{aa} for the modal class is not one.

Formula (6) on p. 740 of [66] should have contained a radical in the denominator, so that the correct formula is

$$T = \sqrt{[\chi^2/\nu]/\sqrt{(\alpha - 1)(\beta - 1)}}.$$

We thank Vernon Davies (Washington State) for calling this to our attention, and we apologize to him and to other readers for an erroneous corrigendum about this point on page 578 of the December 1957 issue of this *Journal*, in which a solidus was missing before the inner radical.

3. WORK ON MEASURES OF ASSOCIATION IN THE LATE NINETEENTH AND EARLY TWENTIETH CENTURIES

3.1. *Doolittle, Peirce, and contemporary Americans; Köppen.* In the 1880's, interest arose in American scientific circles regarding measures of association. Such eminent men as M. H. Doolittle, of Doolittle's method, and C. S. Peirce, the well-known logician and philosopher, took part in the discussion.

Apparently it began with the publication [47] by J. P. Finley, Sergeant, Signal Corps, U.S.A., of his results in attempting to predict tornadoes. During four months of 1884, Finley predicted whether or not one or more tornadoes would occur in each of eighteen areas of the United States. The predictions generally covered certain eight-hour periods of the day. One of Finley's summary tables is given below as an example.

COMPARISON OF FINLEY TORNADO PREDICTIONS AND OCCURRENCES, APRIL, 1884. SOURCE: [47, p. 86]

TABLE SHOWS FREQUENCIES OF TIME PERIOD—GEOGRAPHICAL AREA COMBINATIONS IN EACH CELL

		Occurrence		
		Tornado	No Tornado	Totals
Prediction	Tornado	11	14	25
	No Tornado	3	906	909
	Totals	14	920	934

Thus, for example, in 14 out of the 934 time period-geoegraphical area combinations considered, one or more tornadoes occurred; out of these 14, Finley predicted 11. Since Finley's predictions were correct in 917 out of 934 cases he gave himself a percentage score of 100 (917/934) = 98.18 per cent.* Thus he used the diagonal sum mentioned in Section 8 of [66].

This score, as a measure of association between prediction and occurrence, is wholly inappropriate for Finley's study. A completely ignorant person could always predict "No Tornado" and easily attain scores equal to or greater than Finley's; in the above example, always predicting "No Tornado" would give rise to a score of 100(920/934) = 98.50 per cent. (Of course it is clear that Finley did appreciably better than chance; the question is that of measuring his skill by a single number.)

It was not long before Finley was taken to task. G. K. Gilbert [55] pointed out the fallacy and suggested another procedure, prefacing his suggestion, with commendable humility, in the following words:

> "It is easier to point out an error than to enunciate the truth; and in matters involving the theory of probabilities the wisest are apt to go astray. The following substitute for Mr. Finley's analysis is therefore offered with great diffidence, and subject to correction by competent mathematicians." [55, p. 167]

If Finley's table is written in terms of proportions rather than frequencies, and in the notation of [66], it is of form

		Occurrence		
		Tornado	No Tornado	Total
Prediction	Tornado	ρ_{11}	ρ_{12}	$\rho_{1\cdot}$
	No Tornado	ρ_{21}	ρ_{22}	$\rho_{2\cdot}$
	Total	$\rho_{\cdot 1}$	$\rho_{\cdot 2}$	1

Gilbert suggests that a sensible index of prediction success would be the quantity

* Finley actually obtained such percentage scores for each geographical area separately and then averaged the scores. For April the average was 98.51 per cent.

$$\frac{\rho_{11} - \rho_{1\cdot}\rho_{\cdot 1}}{\rho_{1\cdot} + \rho_{\cdot 1} - \rho_{11} - \rho_{1\cdot}\rho_{\cdot 1}},$$

and he lists a number of formal properties that this index has. For example, it is ≤ 1; it is zero when $\rho_{11} = \rho_{1\cdot}\rho_{\cdot 1}$; it has desirable monotonicities; etc. Finally Gilbert mentions the difficulties of extending his index to prediction problems with more than two alternatives. H. A. Hazen [77] criticized Gilbert's paper, and suggested an alternative index of predictive success based upon a weighted scoring scheme that gave decreasing credit to occurring tornadoes as they fell further from the center of the predicted region.

In the same year that Gilbert's paper appeared, C. S. Peirce [115] suggested a much less ad hoc index of prediction success. Peirce pointed out that one could think of the observed results as obtained by using an infallible predictor a proportion θ of the time, and a completely ignorant predictor the remaining proportion $1-\theta$ of the time. The infallible predictor predicts "Tornado" if and only if a tornado will occur. The ignorant predictor uses an extraneous chance device that precicts "Tornado" with frequency ψ and "No Tornado" with frequency $1-\psi$. Thus what we are asked to contemplate is a mixture of the two 2×2 sets of probabilities

$\rho_{\cdot 1}$	0
0	$\rho_{\cdot 2}$

$\rho_{\cdot 1}\psi$	$\rho_{\cdot 2}\psi$
$\rho_{\cdot 1}(1-\psi)$	$\rho_{\cdot 2}(1-\psi)$

with weights θ and $1-\theta$ respectively. The meanings of the four cells in these tables are the same as in the preceding tables.

The mixed table is, therefore,

$\theta\rho_{\cdot 1}+(1-\theta)\rho_{\cdot 1}\psi$	$(1-\theta)\rho_{\cdot 2}\psi$
$(1-\theta)\rho_{\cdot 1}(1-\psi)$	$\theta\rho_{\cdot 2}+(1-\theta)\rho_{\cdot 2}(1-\psi)$

and Peirce inquires what values of θ and ψ will reproduce the actually observed 2×2 table. (Note that for any θ and ψ the column marginals of the mixed table are $\rho_{\cdot 1}$ and $\rho_{\cdot 2}$.)

For this approach to make sense, θ and ψ must be uniquely defined in terms of the actual ρ_{ab} table. From the (1, 2) cell, we require

$$(1 - \theta)\psi = \rho_{12}/\rho_{\cdot 2},$$

whence from the (1, 1) cell

$$\theta = \frac{\rho_{11}\rho_{22} - \rho_{12}\rho_{21}}{\rho_{\cdot 1}\rho_{\cdot 2}} = \frac{\rho_{11} - \rho_{1\cdot}\rho_{\cdot 1}}{\rho_{\cdot 1}\rho_{\cdot 2}},$$

and

$$\psi = \frac{\rho_{12}\rho_{\cdot 1}}{\rho_{12}\rho_{\cdot 1} + \rho_{21}\rho_{\cdot 2}}.$$

Substitution shows that these values form a unique solution. The only difficulty occurs when θ is negative, for then it can scarcely be a probability. θ itself is suggested as the measure of association in the sense of prediction success. Note that

$$\theta = \frac{\rho_{11}}{\rho_{\cdot 1}} - \frac{\rho_{12}}{\rho_{\cdot 2}},$$

or the difference between the conditional columnwise probabilities of a tornado prediction.

If $\theta=1$, prediction is considered as good as possible, since it is equivalent to infallible prediction. If $\theta=0$, prediction is as poor as it can be without being perverse, since it is equivalent to randomized prediction using the row marginal frequencies of the table under investigation; that is, it corresponds to independence. Further, the θ that makes $\theta\rho_{\cdot 1}+(1-\theta)\rho_{\cdot 1}\psi$ equal to ρ_{11} has an operational interpretation in terms of a hypothetical, if perhaps far-fetched, model of activity. As θ increases, prediction improves.

This proposal by Peirce is of a kind that may be called latent structure measures. We discuss this kind of measure later on in Section 4.9. Peirce's measure, θ, was independently proposed and differently motivated by W. J. Youden in 1950 [66, p. 745, footnote].

Peirce mentions the extension of his approach to larger tables but gives no details. He concludes by suggesting another index that takes into account the "profit, or saving, from predicting a tornado, and . . . the loss from every unfulfilled prediction of a tornado (outlay in preparing for it, etc.). . . ." Thus Peirce, writing in 1884, is the first person of whom we know to discuss the measure of association problem with the intent of giving operationally meaningful measures. Of course, further study might bring earlier proposals to light.

Very soon after Peirce's letter appeared, M. H. Doolittle [35] discussed the topic at the December 3, 1884, meeting of the Mathematical Section of the Philosophical Society of Washington. Doolittle argued for a symmetrized version of Peirce's index, suggesting on rather ad hoc grounds the product of the two possible asymmetrical Peirce quantities

$$\frac{\rho_{11}\rho_{22} - \rho_{12}\rho_{21}}{\rho_{\cdot 1}\rho_{\cdot 2}}, \quad \frac{\rho_{11}\rho_{22} - \rho_{12}\rho_{21}}{\rho_{1\cdot}\rho_{2\cdot}}.$$

This product is simply the mean square contingency, and may be the first occurrence of this chi-square-like index. Doolittle also alluded to the difficulty of extending such measures beyond the 2×2 case.

At a subsequent meeting of the Mathematical Section (February 16, 1887), Doolittle [36] continued his discussion in more general terms than those of measures of prediction success alone. His discussion is similar at points to that of Yule's 1900 paper [149] and he attempts to develop a rationale for the quantity we call the mean square contingency; Doolittle called it the discriminate association ratio. At a third meeting (May 25, 1887), Doolittle [36] concluded his discussion with a criticism of Gilbert's criticism of Finley.

We cannot forbear presenting a quotation from Doolittle in which he struggles to state verbally the general approach he favors.

"The general problem may be stated as follows: Having given the number of instances respectively in which things are both thus and so, in which they are thus but not so, in which they are so but not thus, and in which they are neither thus nor so, it is required to eliminate the general quantitative relativity inhering in the mere thingness of the things, and to determine the special quantitative relativity subsisting between the thusness and the soness of the things." [36, p. 85]

What is a reasonable measure of prediction success for Finley's tables in terms of our λ measures? In this case, λ_b is zero, reflecting the fact that knowledge of Finley's prediction would be no better than ignorance of it in predicting a tornado. If, however, we adjust Finley's table so that the column marginals are equal, while conditional column frequencies remain unchanged, we obtain $\lambda_b{}^* = .67$. This means that if Finely's prediction method were used in a world in which tornadoes occur half the time, we could reduce the error of prediction 67% by knowing Finley's prediction as against not knowing it. We might go further and make both column and row marginals equal, obtaining $\lambda_b{}^* = .88$. The interpretation of this is the same as before except that now Finley is allowed to use the knowledge that tornadoes occur half the time, so that he will predict a tornado half the time.

It may, of course, be cogently argued that in situations such as Finley's it is misleading to search for a single numerical measure of predictive success; and that rather the whole 2×2 table should be considered, or at least two numbers from it, the proportions of false positives and false negatives.

We conclude this section by mentioning briefly some suggestions made by German meteorologists at about the same time. As early as 1870, W. Köppen had considered association measures in connection with his study of the tendency of meteorological phenomena to stay fixed over time. This is related to the problem of measuring prediction, although it is not quite the same. Köppen's basic article on the topic appears to be [91]; an exposition is given by H. Meyer [108, Chapters 11 and 13] together with further references. Köppen and Meyer discuss the question of measuring constancy in various contexts; one relates to a 2×2 table with both classifications the same but referring to different times, and with the two marginal pairs of frequencies the same. For example, the table might be of the following form:

Wind at 2 P.M. at an
observation station

		North	Not North	
Wind at preceding 8 A.M. at	North	ρ_{NN}	$\rho_{N\bar{N}}$	ρ_N
the observation station	Not North	$\rho_{\bar{N}N} = \rho_{N\bar{N}}$	$\rho_{\bar{N}\bar{N}}$	$1 - \rho_N$
		ρ_N	$1 - \rho_N$	1

In this case Köppen (as we interpret his discussion) suggests measuring constancy of wind direction between 8 A.M. and 2 P.M., with respect to the dichotomy North vs. Not North, by

$$\frac{\rho_N(1 - \rho_N) - \rho_{N\bar{N}}}{\rho_N(1 - \rho_N)},$$

or the difference between the probability of a change from North under independence and the same actual probability, this difference taken relative to the probability under independence.

In 1884, an article either by Köppen or someone probably influenced by him [92] suggested a measure of reliability between meteorological prediction and later occurrence in the 3×3 ordered case. The measure was

$$\sum \rho_{aa} + \frac{1}{2} \sum_{|a-b|=1} \sum \rho_{ab},$$

as in Section 8.3 of [66]. The next year, H. J. Klein [89] discussed the simpler measure $\sum \rho_{aa}$ in the general $\alpha \times \alpha$ reliability case.

Bleeker [10] presents an analytical survey of the above early American and German suggestions in the field of meteorological prediction, together with a discussion of many other papers. In Section 4.10 of this paper we survey more recent uses of association measures in meteorology.

3.2. *Körösy, Jordan, and Quetelet.* In [85], Charles Jordan discusses measures of association introduced by József Körösy in the late nineteenth century. Körösy wrote extensively on the effectiveness of smallpox vaccination, and he was led to introduce various measures of association for 2×2 tables in order to summarize and interpret his large quantities of data. Among the several measures discussed by Körösy for 2×2 tables, at least one is equivalent to Yule's Q and hence to our γ (see [66].)

Jordan [85] extends one of Körösy's measures to $\alpha \times \beta$ tables. In our notation, the extended measure is found as follows. For a 2×2 table, Körösy had proposed $(\rho_{11}\rho_{22})/(\rho_{12}\rho_{21})$ as a natural measure of association. Jordan suggests forming all possible $\alpha\beta$ pooled 2×2 tables out of an $\alpha \times \beta$ table, each of form

ρ_{ab}	$\rho_{a\cdot} - \rho_{ab}$
$\rho_{\cdot b} - \rho_{ab}$	$1 - \rho_{a\cdot} - \rho_{\cdot b} + \rho_{ab}$

and averaging the corresponding 2×2 measures to obtain an over-all measure.

Jordan states in [85] the maximum value for the mean square contingency coefficient, ϕ^2. (Jordan also gives this maximum value in another related paper, [86]. The same maximum value has also been given by Cramér [66, p. 740].) Jordan further discusses Körösy's proof and use of the fact that, if in a 2×2 table we observe only a proportion of individuals in a column (i.e., if there is a probability of selection), then, providing the selection probabilities in the two cells of the column are equal, Yule's Q and Körösy's equivalent measure are unaffected. This property of Q is emphasized by Yule in [149] and [150]. Finally, Jordan asserts priority for Körösy's work in the following terms: "Le mérite de Körösy consiste a avòir introduit et utilisé on 1887, c.-à-d. avant l'avènement de la Statistique Mathématique, des grandeurs, mesurant l'association, en bon accord avec les coefficients δ et Q de Yule et ϕ^2 de Pearson utilisés aujourd'hui."

Körösy's writings are not readily available, and we have consulted only one of them [93]. This is a very interesting and sophisticated discussion of statistical material on the efficacy of smallpox vaccination, in which Körösy uses extensively 2×2 table coefficients of association. Emphasis is on the interpretation of such material and on the many ways in which vaccination and smallpox statistics might be consciously or unconsciously distorted, falsified, and biased. (On page 221 of the same volume in which [93] appears, there begins a fascinating discussion of a case of falsification of smallpox-vaccination data. The culprit was an anti-vaccinationist, and the detective work was done by Körösy.)

The question of priority in the use of simple measures of association for 2×2 tables scarcely seems very important. However, it may be of historical interest to note that Yule, in his first (1900) paper on the subject [149] speaks of Quetelet's use of a measure of association in 2×2 tables: $(\rho_{11}-\rho_{1\cdot}\rho_{\cdot 1})/(\rho_{1\cdot}\rho_{\cdot 1})$, in our notation.[1] In fact, Yule named his coefficient "Q" after Quetelet [150, p. 586]. The work by Quetelet of which Yule writes is not accessible to us, but in another place [119] Quetelet uses another very natural measure for comparing (say) the two rows of a 2×2 table in a case wherein they correspond to two binomial populations. He simply takes the ratio of the two binomial p's: $(\rho_{11}/\rho_{1\cdot})/(\rho_{21}/\rho_{2\cdot})$. This ratio probably has been used since nearly the beginning of arithmetic. Of course, neither of the two measures last mentioned have the symmetry of chi-square or of Yule's Q, so that perhaps Jordan would say that they are not "en bon accord" with the measures of Yule and Pearson.

Biographical, bibliographical, and appreciative material on Körösy may be found in a book by Saile [121] and in an obituary by Thirring [134]. A more recent paper by Jordan on the general question of association measures is [87].

3.3. *Benini.* In 1901, the Italian demographer and statistician, R. Benini [4, pp. 129 ff.] suggested measures of attraction and repulsion for 2×2 tables in which the categories of the two dichotomies were the same, or closely related. Benini was mainly concerned at this time with the association between dichotomous characteristics of husband and wife among married couples, for example the association between categories of civil status. Among marriages in Italy during 1898, Benini gives the following 2×2 breakdown of premarital civil status (in relative frequencies):

		Wife: Unmarried	Wife: Widow	Totals
Husband	Unmarried	.8668	.0275	.8943
	Widower	.0742	.0315	.1057
	Totals	.9410	.0590	1

[1] Note that this is the same as the suggestion by Köppen mentioned in Section 3.1.

Comparing this with the corresponding "chance" table obtained by multiplying marginal frequencies, Benini observed that there clearly was association between the premarital civil statuses of husband and wife. To measure the attraction between similar premarital civil statuses, he suggested the following measure (our notation):

$$\frac{\rho_{11} - \rho_{1\cdot}\rho_{\cdot 1}}{\text{Min}(\rho_{1\cdot}, \rho_{\cdot 1}) - \rho_{1\cdot}\rho_{\cdot 1}} = \frac{\rho_{22} - \rho_{2\cdot}\rho_{\cdot 2}}{\text{Min}(\rho_{2\cdot}, \rho_{\cdot 2}) - \rho_{2\cdot}\rho_{\cdot 2}}$$

on the grounds that, when the numerator is nonnegative, the denominator gives the maximum possible value of the numerator for fixed marginals. The numerator is the usual quantity on which 2×2 measures of association are based. When the numerator is negative, a slight revision of the formula provides Benini's measure of repulsion. In the above example Benini's measure of attraction has the value

$$\frac{.8668 - .8415}{.8943 - .8415} = \frac{253}{528} = .479.$$

In 1928, Benini [5] extended his method of analysis by suggesting a separation of the 2×2 population into two 2×2 subpopulations, one with two cells empty, and the other with all marginal frequencies equal to 1/2. Then his measure of attraction (or repulsion) would be computed only for the second sub-population. This represents one way of eliminating the effect of unequal marginals in comparing several 2×2 populations. (In Section 5.4 of [66] another way of attaining this goal was briefly discussed.) A variation of this point of view, much akin to latent structure analysis (see Section 4.9), was applied by Benini to sex-ratios in twins in order to estimate the fractions of fraternal and identical twins in the population.

Benini's work has been discussed by a number of Italian statisticians. An early discussion was by Bresciani in 1909 [15]. A. Niceforo [110, pp. 383–91] and [111, pp. 462–8] also considers Benini's suggestions, and provides an entertaining discussion, with many examples, of several aspects of cross classifications. We refer in particular to Chapter 16 of [111]. A lengthy critical analysis of Benini's suggestions, as applied to matrimonial association, was given by R. Bachi [3]. Some further articles dealing with Benini's work are those of G. de Meo [31], F. Savorgnan [126], G. Andreoli [2], and C. E. Bonferroni [13]. Benini's first measure of attraction was independently suggested by Jordan [87] in 1941, by H. M. Johnson [84a] in 1945, and by L. C. Cole [28] in 1949. No doubt there have been many other independent suggestions of this measure. It has been frequently used by psychologists and sociologists in recent years and called, descriptively enough, $\phi/\phi_{\max}$.

Benini's first measure has recently been critically reviewed by D. V. Glass, J. R. Hall, and R. Mukherjee [63a, pp. 195–96, 248–59] in a book by these writers and others on social mobility in Britain. Glass et al. deal mostly with $\alpha\times\alpha$ cross classifications of father vs. son occupational status; their general approach is to construct a number of 2×2 condensed cross classifications from a larger $\alpha\times\alpha$ one, with the condensed dichotomies of form father (son) in occupational status a vs. not in status a. Then the 2×2 condensations are examined by looking at three of the ratios $\rho_{ab}/(\rho_{a\cdot}\rho_{\cdot b})$.

3.4. *Lipps*. In 1905, G. F. Lipps [100] discussed various ways of describing dependence in a two-way cross classification. For the 2×2 case, Lipps independently proposed Yule's Q. For larger tables, Lipps points out that $(\alpha-1)\cdot(\beta-1)$ numbers are required to describe the dependence in full; he argues againt use of a single numerical measure of association in these words: "Es ist demzufolge nicht zulässig (ausser wenn $r=s=2$ [$\alpha=\beta=2$ in our notation]) einen einzigen Wert als schlechthin gültiges Mass der Abhängigkeit aufzustellen" [100, p. 12]. However, he refers, in a footnote, to articles on correlation by Galton and K. Pearson in contradistinction.

It is interesting to notice that, in the last section of his paper, Lipps proposed a quantity equivalent to Kendall's rank correlation coefficient, τ. The quantity Lipps proposed is Kendall's $P=n(n-1)(\tau+1)/4$ where n is sample size. Lipps suggested testing for independence by this quantity, and to implement this he computed its mean and variance under the hypothesis of independence. A year later Lipps [101] discussed $2P-\binom{n}{2}=\binom{n}{2}\tau$. Material on Lipps' work, and on other early ranking methods, is presented by Wirth [147, particularly Chapter 4, Section 28]. A discussion of the history of Kendall's τ is given by Kruskal [96].

3.5. *Tönnies*. The German sociologist, F. Tönnies, suggested in 1909 [137] a measure of association for square cross classifications in which both polytomies are ordered. A later paper is [138]. Tönnies presents his measure, which is related to the so-called Spearman foot-rule, in terms of continuous, rather than grouped, variates, but he immediately collects them into groups on the basis of their relative magnitudes.

The measure, in our terminology, is found by first adding all ρ_{aa}'s, i.e. all ρ_{ab}'s in the main diagonal, and multiplying this sum by 2. To this is added the sum of all ρ_{ab}'s in the two diagonals neighboring the main diagonal. Then an analogous weighted sum is computed for the counter-diagonal and its two neighbors, and this is subtracted from the first sum. In terms of a formula, Tönnies looks at

$$\left[2\sum_{a-b=0}\sum\rho_{ab}+\sum_{a-b=\pm 1}\sum\rho_{ab}\right]-\left[2\sum_{a+b-1=\alpha}\sum\rho_{ab}+\sum_{a+b-1=\alpha\pm 1}\sum\rho_{ab}\right].$$

He compares this quantity with $2-(2/\alpha)$, its maximum possible absolute value. Thus Tönnies' measure is of the kind discussed briefly by us in Section 8.3 of [66].

H. Striefler [130] provides an exposition of Tönnies' measure and suggests an extension.

3.6. *Deuchler*. In 1914, the German educational psychologist, Gustav Deuchler [32], continued the earlier work of Lipps (Section 3.4) on the quantity now called Kendall's τ. Deuchler worked on the distribution of τ, both under the null hypothesis of independence and under alternative hypotheses, on methods of computing τ, and on modifications when ties are present.[2]

[2] For further discussion of Deuchler's work on τ itself we refer to [96]. For information about other aspects of Deuchler's work, and for remarks about an unpublished monograph by Deuchler, we refer to [95]. A microfilm of this unpublished manuscript is in our hands, and we will try to make it available on request.

A few years later, Deuchler [33] returned to the question of multiple ties in both coordinates, so that he was really concerned with cross classifications having meaningful order for both polytomies. For this situation Deuchler suggested as a measure of association (in our notation)

$$\mathfrak{R} = \frac{\Pi_{s\leq} - \Pi_{d\leq}}{1 - \Pi_{t(\text{both})}} = \frac{\Pi_s - \Pi_d}{1 - \Pi_{t(\text{both})}}.$$

Here $\Pi_{s\leq}(\Pi_{d\leq})$ is the probability that two randomly chosen individuals from the cross classified population will have their A and B categories similarly (dissimilarly) ordered, with a tie *in one polytomy alone* always counting as similarity (dissimilarity), but a tie in *both* categories—i.e. both individuals in the same cell—not counting in either case. $\Pi_{t(\text{both})}$ is the probability that two randomly chosen individuals fall into the same cell, i.e. are tied in both polytomies. Thus $\mathfrak{R}$ is much like our γ [66, Sec. 6] except that Deuchler has $\Pi_{t(\text{both})}$ where we have Π_t.

Actually Deuchler's presentation is in terms of choosing two individuals at random *without* replacement from a finite cross classified population, whereas we in [66] give an interpretation in terms of random choice with replacement. For $\mathfrak{R}$, one obtains the same value of the measure in either interpretation, while γ changes slightly as one goes from the with-replacement to the without-replacement interpretation.

Deuchler develops his $\mathfrak{R}$ by the same scoring scheme as that later used by Kendall. $\mathfrak{R}$ does not have quite as direct an interpretation as γ, but it possesses one characteristic that γ does not have: $\mathfrak{R}$ is 1 (its maximum value) if and only if at most one ρ_{ab} in each row and column is positive *and* the positive ρ_{ab}'s are all concordant. This last means that, denoting the positive ρ_{ab}'s by $\rho_{a_1b_1}$, $\rho_{a_2b_2}$, $\cdots$, with $a_1<a_2<\cdots$, then $b_1<b_2<\cdots$. The examples on p. 750 of [66] show that this property is not true for γ. Note that $|\mathfrak{R}|\leq|\gamma|$.

Deuchler observes that $\mathfrak{R}$ varies as contiguous categories are pooled and he discusses the magnitude of this effect at length. He also compares his $\mathfrak{R}$ with Spearman's rank correlation coefficient, and with the mean square contingency coefficient in the 2×2 case. The applications that Deuchler has in mind, and for which he uses his measure, are to the association between the grades of school children in two subjects or traits. He discusses briefly the situation in which one wishes to analyze such joint gradings on more than two such characteristics. In [34], Deuchler discusses in more detail the 2×2 case.

3.7. *Gini.* In 1914–1916, Corrado Gini [56, 57, 58, 59, 60] examined in detail many distinctions between relationships within a bivariate distribution, and proposed a great variety of measures of association and disassociation.[3] Examples were given to indicate the circumstances under which the various proposed measures might be appropriate.

Many of Gini's measures of association relate to cases in which the bivariate distribution is quantitative or can easily be made so by the use of relevant ordinal scores. For the qualitative case without ordering among the categories

---

[3] We wish to thank Sebastian Cassarino, Department of Italian, University of California, Berkeley, for his assistance in examining Gini's papers.

(*sconesse* categories), and where both polytomies are the same ($A_a = B_a$), Gini [57] proposed as a measure of association the quantity (in our notation)

$$\frac{\sum \rho_{aa} - \sum \rho_{a\cdot}\rho_{\cdot a}}{\sqrt{(1 - \sum \rho_{a\cdot}^2)(1 - \sum \rho_{\cdot a}^2)}}$$

This is based on a sort of indirect scoring scheme, suggested by divergences of cell frequencies from the corresponding marginal products. In the 2×2 case, the above quantity is the appropriately signed square root of the mean square contingency.

In [58], Gini proposed the following variant of the above measure:

$$\frac{\sum \rho_{aa} - \sum \rho_{a\cdot}\rho_{\cdot a}}{1 - \frac{1}{2}\sum |\rho_{a\cdot} - \rho_{\cdot a}| - \sum \rho_{a\cdot}\rho_{\cdot a}},$$

and a number of other variations were discussed systematically in [58] and [60].

We have not found in Gini's papers operational interpretations of his proposed measures. They all seem to be of a formal nature in which consideration of absolute or quadratic differences, followed by averaging, is taken as reasonable without argument. Special attention is paid to denominators so as to make the indices range between 0 and 1 (or -1 and 1) within appropriate limitations for variation in the joint distribution.

In [57, p. 598], Gini briefly discussed polytomies in which the categories are *cyclically* ordered, as for example the months of the year. This type of polytomy was not discussed by us in [66]. Gini suggested the possibility of a measure of association in this case, but he gave little detail. Ten years later Pietra [116] considered the cyclical case in great detail, and since then other Italian authors have written on this topic.

The measures proposed by Gini have formed the basis of a large literature, mostly in Italian. We now cite several publications outlining and discussing Gini's work in this area. First, Gini himself [60, pp. 1458 ff.] gave a systematic outline of his measures. An exposition in English of some of the Gini material was given by Weida [144], and a more detailed exposition by Pietra in the introduction of [116]. A general article on the work of the Italian school is that of Gini [61]; another, of a critical nature, is by Thionet [132]. (The reader of this last article should also refer to subsequent correspondence by Galvani [54] and Thionet [133].) Two recent expositions by Gini are [62] and [63, Chap. 9].

Some further references to the recent Italian literature appear in Section 4.7. In Section 4.4 a measure proposed by Gini in the $\alpha\times 2$ case is discussed in detail.

4. MORE RECENT PUBLICATIONS

4.1. *Textbook discussions. Guilford, Dornbusch and Schmid, Wallis and Roberts.* In [73], J. P. Guilford discusses association in an $\alpha\times\beta$ table from the viewpoint of optimal prediction in a manner essentially equivalent to that of Guttman (see comment and reference in [66, p. 742]), and to that in which we introduced λ_a and λ_b in [66]. This discussion appears in Chapter 10 of the 1942 edition and is amplified in Chapter 14 of the 1950 edition.

In a recent textbook [37, p. 215], S. M. Dornbusch and C. F. Schmid discuss a "coefficient of relative predictability" for $\alpha\times 2$ tables, their G. It is equal to λ_a for $\alpha\times 2$ tables.

W. A. Wallis and H. V. Roberts present the λ measures and γ in their book [142, Chap. 9]. Their notation corresponds to ours as follows:

Wallis-Roberts	$g_{c\cdot r}$	$g_{r\cdot c}$	g	h	S	D	T
Goodman-Kruskal	λ_b	λ_a	λ	γ	$\frac{n^2}{2}\Pi_s$	$\frac{n^2}{2}\Pi_d$	$\frac{n^2}{2}\Pi_t$

and their discussion is in terms of sample frequencies.

4.2. *Reliability measures.* We describe now some papers on measures of association in the reliability context, that is when both polytomies of a cross classification are the same and refer to two methods of assignment. Other papers that deal with reliability measures will be discussed elsewhere, particularly in Sections 4.9 through 4.12, under other classifications.

Wood. In 1928, K. D. Wood [148] suggested several variations of the kind of measure of association described in Section 8.3 of [66] where reliability for ordered polytomies was discussed. Wood's suggestions related to a 4×4 table with $\rho_{a\cdot}=\rho_{\cdot b}=.25$ for all a and b; they were

$$\sum \rho_{aa}, \quad \sum_{|a-b|\leq 1}\sum \rho_{ab}, \quad \sum \rho_{aa} - \sum_{a+b=5}\sum \rho_{ab}, \quad \text{and} \quad \sum_{|a-b|\leq 1}\sum \rho_{ab} - \sum_{|a-b|\geq 2}\sum \rho_{ab}.$$

Actually, Wood's discussion is in terms of sample analogs, and it is wholly motivated by the desire to find sample functions that approximate well to the sample correlation coefficient. To investigate this he divides a sample into 16 parts via its marginal quartiles, computes the above measures, and compares them with the sample correlation coefficient.

Reuning. H. Reuning [120] has recently suggested a new measure of reliability in the case of ordered polytomies. Reuning compares the actual ρ_{ab} table with the table that would result if (a) there were independence between rows and columns, and (b) the marginal distributions were rectangular—he calls this the case of pure chance; its meaning is that each $\rho_{ab}=1/\alpha^2$. Further, in order to use the natural ordering, Reuning suggests pooling all cells such that $|a-b|$ = constant. There are α cells such that $|a-b|=0$, $2(\alpha-1)$ cells such that $|a-b|=1$, $2(\alpha-2)$ cells such that $|a-b|=2, \cdots$, and 2 cells such that $|a-b|=\alpha-1$, the maximum difference. Thus Reuning is led to compare

$$\begin{array}{rll} \sum \rho_{aa} & \text{with} & \alpha(1/\alpha^2) = 1/\alpha \\ \sum\limits_{|a-b|=1} \rho_{ab} & \text{with} & 2(\alpha-1)/\alpha^2 \\ \sum\limits_{|a-b|=2} \rho_{ab} & \text{with} & 2(\alpha-2)/\alpha^2 \\ & \vdots & \\ \rho_{1\alpha}+\rho_{\alpha 1} & \text{with} & 2/\alpha^2 \end{array}$$

In order to obtain a measure of reliability, Reuning in effect considers the following χ^2-like quantity

$$\sum_{k=0}^{\alpha-1} \frac{\left[\sum_{|a-b|=k} \rho_{ab} - \frac{\text{No. of summands in } \sum_{|a-b|=k}}{\alpha^2} \right]^2}{\frac{\text{No. of summands in } \sum_{|a-b|=k}}{\alpha^2}} .$$

Reuning also considers $\Sigma\rho_{aa}$, a measure mentioned in [66].

The above presentation differs slightly from that given by Reuning, first, because he works with sample instead of population quantities, and, second, because he emphasizes testing rather than estimation. If we regard the population characteristic in the above display as a general measure of reliability (and it is not wholly clear from Reuning's paper whether he so regards it) some problems of interpretation arise, stemming from the comparison with the "pure chance" cross classification. For one thing, if $\Sigma\rho_{aa}=1$, so that reliability in the ordinary sense is perfect, Reuning's measure takes the value $\alpha-1$, which is by no means its maximum possible value. On the other hand, if $\rho_{1\alpha}+\rho_{\alpha 1}=1$, so that reliability in the ordinary sense is about as poor as can be, Reuning's measure takes the value $(\alpha^2-2)/2$, which is actually greater than its value for $\Sigma\rho_{aa}=1$ (unless $\alpha=2$, when the two values are equal).

The "pure chance" or uniform table as a basis of comparison had been put forward by Andreoli [1] in 1934. H. F. Smith [126a] uses the same device of pooling along diagonals as does Reuning, but in the context of a comparative test of two square cross classifications.

Cartwright. D. S. Cartwright, for the case of unordered polytomies, has recently [19] suggested a measure of interreliability when there are two *or more* classifications, each with the same polytomy. He thinks of the common polytomy as possible judgments about members of the population on the part of J judges, so that

$$\rho_{a_1 a_2 \cdots a_J}$$

is that fraction of the population allocated by judge 1 to class a_1, by judge 2 to class a_2, etc., where $a_j=1, 2, \cdots, \alpha$. His measure of reliability, in our notation, may be written as

$$\frac{2}{J(J-1)} \sum_j \sum_{k>j} \sum_{a_j=a_k} \rho_{\cdots a_j \cdots a_k \cdots} ,$$

or the probability that two different randomly chosen judges out of the J judges will allocate a random member of the population to the same class. For $J=2$, this becomes just $\Sigma\rho_{aa}$.

Cartwright's presentation of his measure differs superficially from the above. He also considers distribution theory for the sample analogue of the above measure under special restrictive conditions.

4.3. *Measures that are zero if and only if there is independence.* The traditional χ^2-like measures of association, unlike the λ and γ measures discussed by us in

[66], have the property that they take a particular value, zero, if and only if there is independence in the cross classification, i.e., $\rho_{ab}=\rho_{a\cdot}\rho_{\cdot b}$. This property has seemed important to a number of workers, and they have proposed measures of association with the property but different from the traditional measures. In some cases, other formal properties have also been emphasized. We now discuss several such proposals that do not fit more naturally into other sections of this survey.

So far as we know, none of the measures discussed here have operational interpretations of the kind we have argued for in [66], and indeed this is not surprising. For a measure with an operational interpretation measures, so to speak, one aspect or dimension of association. Hence, if a given cross classification exhibits no association along this aspect or dimension one would expect a zero value for the measure, even if there is association in other senses. That is why we are not troubled by the fact that the λ and γ measures can be zero even though there is dependence. Note that if there is independence the λ and γ measures are zero. This is to be expected, since independence should correspond to lack of association in *any* sense.

Cramér. In 1924, H. Cramér [29] suggested for an $\alpha\times\beta$ table the measure

$$\min \sum_a \sum_b (\rho_{ab} - u_a v_b)^2$$

where the minimum is computed over all numbers $u_1, \cdots, u_\alpha; v_1, \cdots, v_\beta$. This quantity is zero if and only if there is independence, and is always $\leq .25$. It suffers from having no definite value in the case of complete dependence.

Cramér says [29, p. 226] that "... there is no absolutely general measure of the degree of dependence. Every attempt to measure a conception like this by a single number must necessarily contain a certain amount of arbitrariness and suffer from certain inconveniences."

Steffensen. In 1933, J. F. Steffensen [127] proposed the following measure of association for cross classifications:

$$\psi^2 = \sum_a \sum_b \rho_{ab} \frac{(\rho_{ab} - \rho_{a\cdot}\rho_{\cdot b})^2}{\rho_{a\cdot}(1 - \rho_{a\cdot})\rho_{\cdot b}(1 - \rho_{\cdot b})}$$

in our notation. (See Lorey [105] for a discussion.) Apparently Steffensen's motivation was to avoid certain formal inadequacies of previously suggested measures. For example, Steffensen points out that his ψ^2 attains its upper limit of 1 if and only if the two classifications are functionally related, i.e. if and only if exactly one ρ_{ab} in each row and in each column is positive. Steffensen gives no operational interpretation for ψ^2. Note that ψ^2 is an average of all 2×2 mean square contingencies formed from each of the $\alpha\beta$ cells of the cross-classification and its complement; in this it resembles the measure proposed by Jordan [85] that we discussed in Section 3.2.

The next year, Steffensen [128] returned to ψ^2 in greater detail. (In [127] the measure had appeared only in a nonnumbered page of errata, as a better version of a similar measure, given in the article proper, that Steffensen later decided was unsatisfactory.) Then Steffensen suggested a variant,

$$\omega = \frac{2 \overline{\sum} \overline{\sum} (\rho_{ab} - \rho_{a\cdot}\rho_{\cdot b})}{\overline{\sum} \overline{\sum} (\rho_{ab} - \rho_{a\cdot}\rho_{\cdot b}) + 1 - \overline{\sum} \overline{\sum} \rho_{ab}^2},$$

where $\overline{\sum} \overline{\sum}$ means summation over those cells for which $\rho_{ab} > \rho_{a\cdot}\rho_{\cdot b}$. He showed that ω, along with ψ^2, (1) lies between 0 and 1, (2) is 0 if and only if independence obtains, and (3) is 1 if and only if exactly one ρ_{ab} in each row and column is positive. Finally, an extension to the case of continuous bivariate distributions was suggested.

Immediately following [128] an editorial [114] (presumably by Karl Pearson) criticized Steffensen's suggestions with arguments based on the assumption of an underlying continuous distribution. First, the editorial said that the continuous analogue of ψ^2 would be identically zero because of the presence of squared differentials. Then it argued that a measure of association for cross classifications should *not* be able to attain the value unity, because, while complete dependence might exist between the two polytomies, it could well be the case that a finer cross classification would show that *within* the original cells complete association did not exist. These arguments were used to contrast Steffensen's suggestions with the coefficient of mean square contingency, to the latter's favor. The editorial concluded with a numerical comparison of ψ^2 and the coefficient of mean square contingency for a number of artificial cross classifications, and it stated that ψ^2 tends to be too low, with values crowded in the interval [0, .25], even for quite sizable intuitive association.

In 1941, Steffensen [129] returned to his discussion of ω. He presented a natural generalization to the density function case and showed that the three properties mentioned above still essentially held. A lengthy discussion of the generalized ω in the bivariate normal case was given, and the paper concluded with a rebuttal to the arguments of [114].

This discussion reinforces our beliefs that it is essential to give operational interpretations of measures of association and that the mere fact that a measure can range from 0 to 1 (say) is of little or no use in understanding it.

Pollaczek-Geiringer. In 1932 and 1933, Hilda Pollaczek-Geiringer [117, 118], motivated by considerations similar to those adduced by Steffensen, suggested a measure of association for any bivariate distribution, continuous or discrete. The measure may also be applied, as Pollaczek-Geiringer suggested, to a cross-classification in which both polytomies are ordered. In our notation, the suggested measure for this case is

$$\frac{\sum_a \sum_b (A_{ab}D_{ab} - B_{ab}C_{ab})}{\sum_a \sum_b (A_{ab}D_{ab} + B_{ab}C_{ab})},$$

where

$$A_{ab} = \sum_{a' \leq a} \sum_{b' \leq b} \rho_{a'b'} \qquad B_{ab} = \sum_{a' > a} \sum_{b' \leq b} \rho_{a'b'}$$

$$C_{ab} = \sum_{a' \leq a} \sum_{b' > b} \rho_{a'b'} \qquad D_{ab} = \sum_{a' > a} \sum_{b' > b} \rho_{a'b'}$$

Pollaczek-Geiringer gives no operational interpretation. Her measure has a certain similarity to our γ [66, Section 6] especially if it is modified by replacement of the summations with weighted sums, having ρ_{ab}'s as weights.

Höffding. In 1941 and 1942, W. Höffding (now Hoeffding) presented two very interesting papers bearing on measures of association for cross classifications. Höffding's first paper on cross classifications [79] was based on a prior paper of his [78] that had dealt solely with the bivariate density function case. In [78], it was urged that measures of association should be invariant under transformations, monotone in the same direction, of the associated random variables. Several measures having this invariance were presented and their properties discussed. The cross classifications of [79] were considered as arising from underlying density function distributions by rounding. Hence their cumulative distribution functions are only known at points of a rectangular lattice, and their density functions are only known via averages over cells. In order to apply the suggestions of [78], Höffding replaced a cross classification by a density function distribution with constant density within each cell, proportional to its ρ_{ab}. (This might appear to make matters depend on metrics for the two classifications, but any such dependence is a notational artifact, disappearing later because of invariance.) Then Höffding applied to this "step-function" density the measures of [78]. The first was the correlation coefficient between the probability integral transforms of the marginal random variables (this is the so-called grade correlation, or population analogue of Spearman's rank correlation coefficient). Höffding obtained

$$\bar{\rho} = 3 \sum_a \sum_b \rho_{ab} \left[2\left(\sum_{a'<a} \rho_{a'\cdot} \right) + \rho_{a\cdot} - 1 \right] \left[2\left(\sum_{b'<b} \rho_{\cdot b'} \right) + \rho_{\cdot b} - 1 \right].$$

A slight modification gave him the more satisfactory

$$\rho^* = \bar{\rho} / \sqrt{(1 - \sum \rho_{a\cdot}{}^3)(1 - \sum \rho_{\cdot b}{}^3)}.$$

Höffding then discussed the extrema that $\bar{\rho}$ and ρ^* can reach, and their values for 2×2 tables. In the 2×2 case, ρ^{*2} is just the mean square contingency.

Höffding then pointed out that his ρ^* is the same as Student's modification of Spearman's rank correlation coefficient [131], provided that appropriate notational translations are made. The article continued with a discussion of mean square contingency and related coefficients, including one that is a function of the quantities

$$\left(\sum_{a' \le a} \sum_{b' \le b} \rho_{a'b'} \right) - \left(\sum_{a' \le a} \rho_{a'\cdot} \right) \left(\sum_{b' \le b} \rho_{\cdot b'} \right),$$

thus giving a measure of departure from independence as defined in terms of cumulative distributions.

In the later portion of [80], Höffding returned to these questions. He distinguished between those cases in which a continuous distribution is considered as underlying the discrete distribution of interest, and those cases in which the discrete distribution itself is of primary interest. For this second situation he suggested a measure of association by analogy with one for density-function distributions suggested earlier in the article. It is simply

$$\tfrac{1}{2}\sum\sum|\rho_{ab}-\rho_{a\cdot}\rho_{\cdot b}|.$$

A modification was then put forward, namely, division by $1-\overline{\sum}\overline{\sum}\rho_{ab}^2$, where $\overline{\sum}\overline{\sum}$ means summation over those (a, b) such that $\rho_{ab}>\rho_{a\cdot}\rho_{\cdot b}$. The result is simply related to Steffensen's ω.

Eyraud. H. Eyraud [43] suggested for the 2×2 table the measure of association $(\rho_{11}-\rho_{1\cdot}\rho_{\cdot 1})/(\rho_{1\cdot}\rho_{\cdot 1}\rho_{2\cdot}\rho_{\cdot 2})$. He discussed its extreme values, its interpretation, and, briefly, its extension to $\alpha\times\beta$ tables. In addition he considered the $2\times 2\times 2$ case.

Fréchet, Féron. M. Fréchet has discussed measures of association in a series of articles (e.g. [50] and [51]) that deal mostly with cases in which a meaningful metric exists for both polytomies. In some more recent articles, [52] and [53], he has studied the extent to which knowledge of the marginals restricts the probabilities of a cross classification. Fréchet's work discusses the extent to which measures of association satisfy a set of formal criteria such as those mentioned earlier in this section.

In two recent publications, [45] and [46], R. Féron has discussed measures of association, again with emphasis on the case when metrics are present, but with some consideration of the purely qualitative case. Several of the measures described in this section are discussed by Féron.

4.4. *Measures of dissimilarity, especially in the $\alpha\times 2$ case.* In considering an $\alpha\times 2$ cross classification, it is natural to approach the question of association by asking about the degree of dissimilarity between the two conditional multinomial populations in the two columns, when compared row by row. This approach has often been taken in the social sciences when columns refer to a dichotomy of interest (Negro-White, Male-Female, etc.) and rows correspond to places, times, or the like. It is, of course, equivalent to speak of a $2\times\beta$ cross classification by simply interchanging rows and columns.

Gini, Florence, Hoover, Duncan and Duncan, Bogue. A measure of dissimilarity in the $\alpha\times 2$ case that has been proposed a number of times, often in variant forms, is the following:

$$D=\frac{1}{2}\sum_{a=1}^{\alpha}\left|\frac{\rho_{a1}}{\rho_{\cdot 1}}-\frac{\rho_{a2}}{\rho_{\cdot 2}}\right|,$$

or half the sum of absolute differences between corresponding conditional probabilities in the first and second columns. The use of D appears to have been first suggested by C. Gini (see [56], [57], [61a]); some more recent publications about this measure are by P. S. Florence [48], E. M. Hoover [82] and [83], O. D. Duncan and B. Duncan [42], and D. J. Bogue [12].

Since the summation in D, if the absolute value signs were omitted, would be $1-1=0$, we see that

$$\sum_a{}^{+}\left\{\frac{\rho_{a1}}{\rho_{\cdot 1}}-\frac{\rho_{a2}}{\rho_{\cdot 2}}\right\}+\sum_a{}^{-}\left\{\frac{\rho_{a1}}{\rho_{\cdot 1}}-\frac{\rho_{a2}}{\rho_{\cdot 2}}\right\}=0$$

where $\sum_a{}^{+}$ indicates summation over nonnegative values of the summand, and $\sum_a{}^{-}$ indicates summation over negative values of the summand. Thus

$$D = \sum_a^+ \left\{\frac{\rho_{a1}}{\rho_{\cdot 1}} - \frac{\rho_{a2}}{\rho_{\cdot 2}}\right\} = - \sum_a^- \left\{\frac{\rho_{a1}}{\rho_{\cdot 1}} - \frac{\rho_{a2}}{\rho_{\cdot 2}}\right\},$$

and we see that D is the difference between the proportion of the population in column 1 appearing in rows for which $\rho_{a1}/\rho_{\cdot 1} > \rho_{a2}/\rho_{\cdot 2}$ and the proportion of the column 2 population appearing in these rows. A similar verbal statement, with the difference taken in the opposite sense, for rows with $\rho_{a1}/\rho_{\cdot 1} < \rho_{a2}/\rho_{\cdot 2}$, corresponds to the second equality of the above display.

Now suppose we think of redistributing the (conditional) column 1 population among its cells so that it becomes equal to the (conditional) column 2 population. This means moving probability mass from the column 1 cells with $\rho_{a1}/\rho_{\cdot 1} > \rho_{a2}/\rho_{\cdot 2}$ to those with the opposite inequality holding, and clearly the minimum proportion of the column 1 population that we must shift to achieve this goal is D. A similar interpretation may be given in terms of redistributing the column 2 population so that it becomes (conditionally) equal to the column 1 population. After such a redistribution, the two cells in each row would have equal conditional probabilities, each conditional on its fixed column marginals. Also, the proportion of the population in a given row that is in column 1 will be the same for each row. Thus D has a useful operational interpretation for some purposes; for example see [42].

The construction of D suggests an ordering of the rows that may be of substantive interest in some contexts. Rearrange the rows so that the row with maximum $(\rho_{a1}/\rho_{\cdot 1}) - (\rho_{a2}/\rho_{\cdot 2})$ becomes the first row, the row with next largest $(\rho_{a1}/\rho_{\cdot 1}) - (\rho_{a2}/\rho_{\cdot 2})$ becomes the second row, and so on. If there are α_* rows with $\rho_{a1}/\rho_{\cdot 1} \geq \rho_{a2}/\rho_{\cdot 2}$, D may then be expressed as

$$\sum_{a=1}^{\alpha_*} \left\{\frac{\rho_{a1}}{\rho_{\cdot 1}} - \frac{\rho_{a2}}{\rho_{\cdot 2}}\right\} = - \sum_{a=\alpha_*+1}^{\alpha} \left\{\frac{\rho_{a1}}{\rho_{\cdot 1}} - \frac{\rho_{a2}}{\rho_{\cdot 2}}\right\}$$

in terms of the reordered cross classification.

Some other easily obtained expressions for D are

$$D = \sum_{a=1}^{\alpha} \left| \frac{\rho_{a1}}{\rho_{\cdot 1}} - \rho_{a\cdot} \right| \bigg/ [2\rho_{\cdot 2}] = \sum_{a=1}^{\alpha} \left| \frac{\rho_{a2}}{\rho_{\cdot 2}} - \rho_{a\cdot} \right| \bigg/ [2\rho_{\cdot 1}]$$

$$= \sum_{a=1}^{\alpha} \sum_{b=1}^{2} \left| \frac{\rho_{ab}}{\rho_{\cdot b}} - \rho_{a\cdot} \right| \bigg/ [4(1 - \rho_{\cdot b})]$$

$$= \sum_{a=1}^{\alpha} \sum_{b=1}^{2} \left| \rho_{ab} - \rho_{a\cdot}\rho_{\cdot b} \right| / [4\rho_{\cdot 1}\rho_{\cdot 2}].$$

The first three of these describe D in terms of absolute differences of form $(\rho_{ab}/\rho_{\cdot b}) - \rho_{a\cdot}$, while the last describes D in terms of the most conventional measure of deviations from cell independence, $\rho_{ab} - \rho_{a\cdot}\rho_{\cdot b}$. This last expression for D resembles the traditional χ^2 kind of measure, but differs from such measures in that the absolute differences are used rather than the squared differences, and the weightings of the terms are different.

Still another mode of description for D may be given in terms of absolute differences between the column conditional probabilities, $\rho_{ab}/\rho_{a\cdot}$, and the column marginals, $\rho_{\cdot b}$. It is easily checked that

$$D = \sum_{a=1}^{\alpha} \left| \frac{\rho_{a1}}{\rho_{a\cdot}} - \rho_{\cdot 1} \right| \rho_{a\cdot}/[2\rho_{\cdot 1}\rho_{\cdot 2}]$$

$$= \sum_{a=1}^{\alpha} \left| \frac{\rho_{a2}}{\rho_{a\cdot}} - \rho_{\cdot 2} \right| \rho_{a\cdot}/[2\rho_{\cdot 1}\rho_{\cdot 2}]$$

$$= \sum_{a=1}^{\alpha} \sum_{b=1}^{2} \left| \frac{\rho_{ab}}{\rho_{a\cdot}} - \rho_{\cdot b} \right| \rho_{a\cdot}/[4\rho_{\cdot 1}\rho_{\cdot 2}].$$

The traditional χ^2-like measures may, of course, also be expressed in analogous equivalent ways in the special case of two columns. For example, $\phi^2 = \chi^2/\nu$ may be expressed as

$$\sum_{a=1}^{\alpha} \sum_{b=1}^{2} (\rho_{ab} - \rho_{a\cdot}\rho_{\cdot b})^2/\rho_{a\cdot}\rho_{\cdot b} = \sum\sum \left(\frac{\rho_{ab}}{\rho_{\cdot b}} - \rho_{a\cdot} \right)^2 \rho_{\cdot b}/\rho_{a\cdot}$$

$$= \sum_{a=1}^{\alpha} \left(\frac{\rho_{a1}}{\rho_{\cdot 1}} - \frac{\rho_{a2}}{\rho_{\cdot 2}} \right)^2 \frac{\rho_{\cdot 1}\rho_{\cdot 2}}{\rho_{a\cdot}}$$

$$= \sum\sum \left(\frac{\rho_{ab}}{\rho_{a\cdot}} - \rho_{\cdot b} \right)^2 \rho_{a\cdot}/\rho_{\cdot b}$$

$$= \sum_{a=1}^{\alpha} \left(\frac{\rho_{a1}}{\rho_{a\cdot}} - \rho_{\cdot 1} \right)^2 \rho_{a\cdot}/\rho_{\cdot 1}\rho_{\cdot 2}$$

$$= 1 - \frac{1}{\rho_{\cdot 1}\rho_{\cdot 2}} \sum_{a=1}^{\alpha} \frac{\rho_{a1}\rho_{a2}}{\rho_{a\cdot}}.$$

The possibilities of expressing a measure in terms of the deviation of ρ_{ab} from $\rho_{a\cdot}\rho_{\cdot b}$, in terms of the deviation of $\rho_{a1}/\rho_{\cdot 1}$ from $\rho_{a2}/\rho_{\cdot 2}$, or in terms of the deviation of $\rho_{ab}/\rho_{a\cdot}$ from $\rho_{\cdot b}$, etc., may give added insight into the nature of the measure by suggesting interpretations and approaches to it from different directions. On the other hand, the same possibility of variant expression may cause confusion in communication and may mislead authors to think that symbolically different expressions correspond to different measures, when in fact the measures are the same. Duncan and Duncan [42] and J. Williams [145] discuss a number of articles where this difficulty seems to exist. The last form given above for ϕ^2 has been discussed by E. Katz and P. Lazarsfeld [87a, p. 373].

Measures of association for the $\alpha \times \beta$ case may be based on the idea of dissimilarity between two columns by averaging in some way the $\beta(\beta-1)/2$ possible values of an $\alpha \times 2$ measure of dissimilarity obtained from pairs of columns in the larger cross classification. Alternatively, one might average the β values of an $\alpha \times 2$ measure obtained by comparing each column of the $\alpha \times \beta$ table with the column of row marginals, $\rho_{a\cdot}$. This approach has been used by Gini and by Fréchet, in references cited earlier.

Boas. In 1922, Franz Boas [11, pp. 432–4] suggested a measure of dissimilarity between one specific column of a cross classification and the column of row marginals, that is between one multinomial population and the (weighted) average of a group of multinomial populations to which the one in question belongs. Boas's suggestion, in our notation, seems to be the following:

Suppose that an individual is chosen at random from the bth column of a cross classified population in accordance with the conditional distribution for that column. That is, the individual falls in the (a, b) cell with probability $\rho_{ab}/\rho_{\cdot b}$. Now suppose that we are told the row in which he falls but *not* told that he came from the bth column. If we guess his column, based on knowledge of his row, in a random manner reproducing the population (as discussed in Section 9 of [66]), we shall guess column b with conditional probability $\rho_{ab}/\rho_{a\cdot}$, where a is the row in which he has fallen. Thus the probability of correctly guessing the cell in which the individual falls, when (i) he is in fact drawn from the bth column, and (ii) we guess his column, knowing only his row, in a random manner reproducing the population, is

$$\sum_a \left(\frac{\rho_{ab}}{\rho_{\cdot b}}\right)\left(\frac{\rho_{ab}}{\rho_{a\cdot}}\right) = \sum_a \rho_{ab}^2/(\rho_{a\cdot}\rho_{\cdot b}).$$

That is, as we undertand it, Boas's measure of dissimilarity between column b and the column of row marginals.

Boas also considers the possibility of changing the table so that it has equal column marginals (see Section 5.4 of [66]).

Long and Loevinger. In working with psychological tests made up of yes-no questions, one may wish to consider association between a particular question and the whole test. This situation may be viewed in the framework of an $\alpha\times 2$ table in which the columns refer to the two possible responses and the rows make up an ordered classification based on the whole test. The ρ_{ab}'s are the proportions of individuals in the population falling into one of the whole-test score classes and responding to the individual question in one of the two possible ways. For this special psychometric situation, measures of association have been proposed and discussed by Long [105] and by Loevinger [102, Chap. 5] and [103].

4.5. *Measures based on Lorenz or cost-utility curves*. For the $\alpha\times 2$ cross classification, where the α rows have a meaningful order (determined from the cross classification itself, as discussed in Section 4.4, or determined from external considerations) the following approach has been suggested. Consider the partial sums

$$X_a = \sum_{i=1}^{a} \frac{\rho_{i1}}{\rho_{\cdot 1}} \quad \text{and} \quad Y_a = \sum_{i=1}^{a} \frac{\rho_{i2}}{\rho_{\cdot 2}},$$

and consider the points (X_a, Y_a) for $a=1, \cdots, \alpha$ in the unit square. The underlying thought is that these are points on a smooth curve expressing a functional relationship between X and Y, but that we only know this curve at the α points (X_a, Y_a). If there is independence in the cross classification, then $Y_a = X_a$ for each a; i.e., the points (X_a, Y_a) lie on the straight line segment going diagonally from (0, 0) to (1, 1). But if there is association, the general shape of the underlying curve suggested by the (X_a, Y_a)'s, and its "distance" from the diagonal line, will describe it. Several measures of association, based on this idea, have been suggested in the literature (see, e.g., [42], [65], [6], [41]), but we shall not discuss them here. In some cases, a structural assumption or smoothing procedure (e.g., the use of straight line segments) is used to obtain a curve from the α points.

4.6. *Measures based on Shannon-Wiener information. McGill, Holloway, Woodbury, Wahl, Linfoot, Halphen.* Some time ago it was suggested to us by J. W. Tukey that measures of association based on the Shannon-Wiener information function might be useful. Since we were unable to satisfy ourselves that such measures would have reasonable interpretations for many contexts in which cross classifications appear, we did not discuss the possibility in [66]. We wish, however, to mention here a few papers in which the information concept is used as the basis of measures of association, although we continue to reserve our opinions about the utility of these proposals outside the area of communication theory.

Perhaps the first such paper is by W. J. McGill [107]. Soon after it, the approach was suggested in a meteorological setting by J. L. Holloway, Jr. and M. A. Woodbury [81]. E. W. Wahl [141] summarizes some of the material of [81]. The measure has been used in meteorology, notably by I. I. Gringorten and his colleagues, [72] and [69]. Two quite recent papers on this general theme are by E. H. Linfoot [99] and E. Halphen [73a].

4.7. *Recent proposals by Italian authors other than Gini.* We have already discussed the early suggestions of Benini (Section 3.3) and the extensive publications by Gini (Section 3.7). Since then, the Italian statistical literature has been replete with articles about one aspect or another of the measurement of association. Nearly all of this literature has been derivative from Gini's 1914–16 publications; the interested reader can find some key references in Section 3.7. We shall not attempt to give a complete outline of this literature, but some of the more interesting articles that have come to our attention will now be listed.

Salvemini. A prolific writer on the theme of measures of association has been T. Salvemini. In [122], he surveyed parts of the field, and suggested some new expressions for Gini's measures in the asymmetrical and unordered qualitative case. In [123], Salvemini discussed the calculation and application of measures of association; the case in which one polytomy is ordered, while the other is not, received consideration. More recently, he has presented [125] an extensive discussion of the whole field of measures of association. References to many other papers by Salvemini may be found in the three articles cited above.

Bonferroni and Brambilla. C. E. Bonferroni has given [12a] a detailed discussion of a number of measures of association, emphasizing relations between the ρ_{ab}'s, $\rho_{a\cdot}$'s and $\rho_{\cdot b}$'s, and pointing out problems and concepts that arise in the three-way cross classification. Another article by Bonferroni in this area is [13]. Closely associated is the work of F. Brambilla [14] who presented a systemic discussion of the field giving particular emphasis to the effects of holding marginals fixed or not and to three-way cross classifications.

Faleschini. Particularly interesting for us is an article by L. Faleschini [44]. His approach is to consider the most probable cell in the bth column, and to compare its conditional probability with some kind of average of the column conditional probabilities in the same row. Thus, if $a^*(b)$ is defined by

$$\rho_{a^*(b)b} \geq \rho_{ab} \qquad \text{(all } a\text{)},$$

Faleschini considers the differences

$$D_b = \frac{\rho_{a^*(b)b}}{\rho_{\cdot b}} - \text{some average of} \frac{\rho_{a^*(b)b'}}{\rho_{\cdot b'}} \qquad (b' = 1, \cdots, \beta).$$

Finally the D_b's are averaged in some way. Thus two averages can be rather arbitrarily introduced. If in the first (that of the conditional probabilities) we weight by $\rho_{\cdot b'}$ $(b' \neq b)$ and 0 $(b' = b)$, and if in the second (that of the D_b's) we weight by $\rho_{\cdot b}$, we obtain, following Faleschini,

$$\sum_b \frac{\rho_{a^*(b)b} - \rho_{a^*(b)\cdot}\rho_{\cdot b}}{1 - \rho_{\cdot b}}.$$

Faleschini appears to feel that this kind of measure should only be used when $\rho_{a^*(b)b}/\rho_{\cdot b} \geq \rho_{a^*(b)b'}/\rho_{\cdot b'}$ for each b and b', but we are not wholly clear about his intent. One difficulty with Faleschini's suggestion is that of interpreting averages of conditional probabilities. Nonetheless, Faleschini's discussion [44] is in terms of a probability model, the drawing of colored balls from urns.

Andreoli. Finally, we wish to mention two articles by G. Andreoli, [1] and [2]. Among the topics discussed is that of association between characteristics of one individual and a *group* of individuals, for example between occupation of father and occupations of his *several* sons.

4.8. *Problems of inference discussed by Wilson, Berkson, and Mainland.* We should like to call attention to three papers in the medical literature that are of interest in connection with measures of association, especially with respect to the very difficult problem of inference from one population to another.

The first is by E. B. Wilson [146]. Wilson emphasizes the importance of specifying the population carefully. For example, consider the 2×2 table

	Dead with evidence of cancer	Not [dead with evidence of cancer]
Dead with evidence of tuberculosis		
Not [dead with evidence of tuberculosis]		

If this table is filled in from the data of a large number of autopsies (so that all individuals represented in the table are dead) one may obtain a very different picture than if the table is filled in from the entire population, alive at a given time and observed one year later.

The second paper is by Joseph Berkson [7]. It considers examples like the above with emphasis on differential selection as a cause of confusion. Berkson proposes a specific mechanism for differential selection in the case of one study of the relation between smoking and lung cancer.

The third paper is by Donald Mainland [106]. He gives in considerable detail an example showing how differential selection can lead to a grossly fallacious inference.

4.9. *Measures based on latent structures.* We have already discussed the 2×2 case measures of association based on latent structures that have been suggested by Peirce (Section 3.1), and Benini (Section 3.3). Both authors suggested that the observable 2×2 cross classification might be regarded as an average or mixture of two or more underlying cross classifications having special characteristics, e.g., independence. The underlying cross classifications are those of the latent classes. One may then take as a measure of association a numerical characteristic of the latent class probabilities together with the averaging or mixing weights, provided that this characteristic is expressible as a function of the four probabilities in the observable cross classification. The latent class structure, which may be considered as either real or fanciful, then provides an interpretation for the proposed measure of association.

Lazarsfeld and Kendall. More recently, Paul F. Lazarsfeld has written extensively about latent class structures; it was indeed Lazarsfeld who introduced the term "latent structure." Although much of Lazarsfeld's work on latent structures has been concerned with much broader problems, he and Patricia Kendall [88, Appendix A] have discussed measures of association based on latent classes in the 2×2 case. We describe first their "index of turnover."

The sort of 2×2 cross classification that Lazarsfeld and Kendall discuss might result from asking people the same yes-or-no question at two different times. The supposed latent structure is that there are really two classes of people in the population of interest, those whose latent attitude towards the question is "Yes," in proportion K_1, and those whose latent attitude is "No," in proportion $K_2=1-K_1$. The actual answers that people give do not, however, always express their latent attitudes, since they may be temporarily swayed in the other direction, may misunderstand, and so on. Suppose that the "Yes" people answer "No" with probability x, and that the "No" people answer "Yes" with probability y. Responses are supposed independent for the people in a given class. Further, in order that the latent structure make sense, we suppose that x and y are $\leq\frac{1}{2}$.

If, now, we choose at random a member of the population, the following four probabilities, arranged in 2×2 form, describe the distribution of his two responses:

		Second answer: Yes	Second answer: No	Totals
First Answer	Yes	$\rho_{11}=K_1(1-x)^2+K_2y^2$	$\rho_{12}=K_1(1-x)x+K_2y(1-y)$	$\rho_{1\cdot}=K_1(1-x)+K_2y$
	No	$\rho_{21}=K_1x(1-x)+K_2(1-y)y$	$\rho_{22}=K_1x^2+K_2(1-y)^2$	$\rho_{2\cdot}=K_1x+K_2(1-y)$
	Totals	$\rho_{\cdot 1}=K_1(1-x)+K_2y$	$\rho_{\cdot 2}=K_1x+K_2(1-y)$	1

This is the observable 2×2 cross classification. Following our general approach, we suppose it known and postpone discussion of sampling problems. Note that $\rho_{12}=\rho_{21}$ and that $\rho_{i\cdot}=\rho_{\cdot i}$ $(i=1, 2)$. There are two independent probabilities among the four of the 2×2 table, and three independent parameters of the latent structure, so one cannot hope to express these parameters in terms of the probabilities. If, however, one assumes that $x=y$, i.e., that the probability of a deviant response is the same for both the "Yes" and "No" latent classes, then the core of the above table simplifies to

$\rho_{11}=x^2-2K_1x+K_1$	$\rho_{12}=x(1-x)$	$\rho_{1\cdot}=K_1(1-2x)+x$
$\rho_{21}=x(1-x)$	$\rho_{22}=x^2-2x(1-K_1)+(1-K_1)$	$\rho_{2\cdot}=-K_1(1-2x)+1-x$

Hence $x=\frac{1}{2}[1\pm\sqrt{1-4\rho_{12}}]$. Since we have assumed $x\leq\frac{1}{2}$, the minus sign should be chosen. Thus $x=\frac{1}{2}[1-\sqrt{1-4\rho_{12}}]$ measures an aspect of association that has a real interpretation in the context of the stated latent structure, since x is the probability of a deviant response. Also $2x(1-x)$ is the probability that a random person answers the question differently at the two times; whence the descriptive term "turnover." And $1-2x(1-x)$ is the probability that a random person answers the question similarly.

One can also easily express K_1 in terms of the ρ's, since

$$K_1 = (\rho_{1\cdot} - x)/(1-2x)$$
$$= \frac{1}{2} + \frac{2\rho_{1\cdot} - 1}{2\sqrt{1-4\rho_{12}}}\,.$$

Further, independence obtains if and only if either $K_1=0$ or 1, or $x=\frac{1}{2}$. Thus x measures an aspect of association, unless $K_1=0$, 1.

A serious difficulty with the above latent structure is that it places severe limitations on the ρ's; only a limited set of 2×2 cross classifications can be fit by it. In fact, it is necessary and sufficient that

$$(1)\ \rho_{12} \leq \tfrac{1}{4}, \qquad (2)\ \rho_{21} = \rho_{21}, \qquad \text{and} \qquad (3)\ \rho_{11} \geq \rho_{1\cdot}\rho_{\cdot1}$$

for a 2×2 cross classification to be describable in terms of the above latent structure.

Kendall and Lazarsfeld also discuss a more general measure, appropriate to some cases in which $\rho_{12}\neq\rho_{21}$, by enlarging the model to embrace three, rather than two, latent classes with special characteristics. In order to exemplify the possibilities, we should like to suggest a new measure that may be more appropriate to some cases in which $\rho_{12}\neq\rho_{21}$. Which measures to use, if any, depends of course on context. The measure we shall now describe might be appropriate when two closely related questions are both asked once, rather than when the same question is asked twice, and we describe it in these terms.

Suppose that on question 1 people give deviant answers (e.g. a "yes" person answers "no") with probability $x_1\leq\frac{1}{2}$, and that on question 2 they give deviant answers with probability $x_2\leq\frac{1}{2}$. The probabilities of deviant response do not depend on the class to which a person belongs. In all other respects the latent structure is the same as before. We then have three independent parameters, K_1, x_1, and x_2 for describing our structure, and the 2×2 table becomes

		Answer to question 2: Yes	No	Totals
Answer to question 1	Yes	$\rho_{11}=K_1(1-x_1)(1-x_2)+K_2x_1x_2$	$\rho_{12}=K_1(1-x_1)x_2+K_2x_1(1-x_2)$	$\rho_{1\cdot}=K_1(1-x_1)+K_2x_1$
	No	$\rho_{21}=K_1x_1(1-x_2)+K_2(1-x_1)x_2$	$\rho_{22}=K_1x_1x_2+K_2(1-x_1)(1-x_2)$	$\rho_{2\cdot}=K_1x_1+K_2(1-x_1)$
	Totals	$\rho_{\cdot1}=K_1(1-x_2)+K_2x_2$	$\rho_{\cdot2}=K_1x_2+K_2(1-x_2)$	1

We may now express K_1, x_1, and x_2 in terms of the ρ's; and x_1 and x_2—thus expressed—are interpretable measures of association in terms of the supposed latent structure. They are the probabilities of deviant responses to the two questions. In order to get a single measure, one might take the average of x_1 and x_2; that is, the probability of deviant response to one of the two questions, which one to be decided by the toss of a fair coin. Or one might use $x_1x_2 + (1-x_1)(1-x_2)$, the probability that a random person answers the two questions similarly.

It is easily seen from the above table that

$$x_1 = \frac{\rho_{1\cdot} - K_1}{1 - 2K_1}, \qquad x_2 = \frac{\rho_{\cdot 1} - K_1}{1 - 2K_1}$$

and that

$$\frac{\rho_{11} - \rho_{1\cdot}\rho_{\cdot 1}}{1 - 2(\rho_{12} + \rho_{21})} = K_1(1 - K_1) = R \quad \text{(say)}.$$

Hence

$$K_1 = \tfrac{1}{2}[1 \pm \sqrt{1 - 4R}]$$

and we see that, for our latent structure to hold, R, as a function of the ρ's, must be $\leq \frac{1}{4}$. Substituting in the above expressions for x_1 and x_2, we obtain

$$x_1 = \frac{1}{2} \mp \frac{2\rho_{1\cdot} - 1}{2\sqrt{1 - 4R}}$$

$$x_2 = \frac{1}{2} \mp \frac{2\rho_{\cdot 1} - 1}{2\sqrt{1 - 4R}}$$

There remains the question about sign choice in the solution of the quadratic for K_1. We want to be able to make the same choice for both x_1 and x_2 so that x_1 and x_2 are $\leq \frac{1}{2}$. This means that $\rho_{1\cdot} - \frac{1}{2}$ and $\rho_{\cdot 1} - \frac{1}{2}$ must have the same sign in the sense that $(\rho_{1\cdot} - \frac{1}{2})(\rho_{\cdot 1} - \frac{1}{2}) \geq 0$. The necessary conditions thus far suggested come to (1) $\rho_{12} + \rho_{21} \leq \frac{1}{2}$, (2) $(\rho_{1\cdot} - \frac{1}{2})(\rho_{\cdot 1} - \frac{1}{2}) \geq 0$, and (3) $\rho_{11} \geq \rho_{1\cdot}\rho_{\cdot 1}$.

Note that if $\rho_{12} = \rho_{21}$, then $\rho_{1\cdot} = \rho_{\cdot 1}$, $x_1 = x_2$, and

$$1 - 4R = \frac{1 - 4\rho_{12} - 4\rho_{11} + 4\rho_{1\cdot}^2}{1 - 4\rho_{12}} = \frac{(1 - 2\rho_{1\cdot})^2}{1 - 4\rho_{12}}.$$

Hence

$$x_1 = x_2 = \frac{1}{2} \mp \frac{2\rho_{1\cdot} - 1}{2(1 - 2\rho_{1\cdot})}\sqrt{1 - 4\rho_{12}} = \frac{1}{2}\,[1 \pm \sqrt{1 - 4\rho_{12}}]$$

and the minus sign must be chosen, to obtain the same result as in the earlier structure. So the structure now being discussed does generalize the earlier one, giving us two turnover indexes.

For the present structure, independence obtains if and only if K_1 is 0 or 1, or if either x_1 or $x_2 = \frac{1}{2}$. Thus we see again that x_1 and x_2 measure aspects of association, unless $K_1 = 0, 1$.

Necessary and sufficient conditions for the present structure to be possible may be expressed in various ways. One such set of conditions is the following pair:

$$(1) \qquad 0 \leq \frac{\rho_{11} - \rho_{1\cdot}\rho_{\cdot 1}}{1 - 2(\rho_{12} + \rho_{21})} \leq \text{Min}\ [\rho_{1\cdot}\rho_{2\cdot},\ \rho_{\cdot 1}\rho_{\cdot 2}]$$

$$(2) \qquad (\rho_{1\cdot} - \tfrac{1}{2})(\rho_{\cdot 1} - \tfrac{1}{2}) \geq 0.$$

4.10. *More recent work on measures of association in meteorology. Gringorten, Bleeker, Brier, and others.* In Section 3.1, we discussed measures of association suggested by Peirce, Doolittle, Köppen and others for meteorological problems. Meteorologists have of course long been interested in the accuracy of weather forecasts, and they have suggested many measures of association between the predicted weather and the weather that actually occurred.

We shall not attempt to survey the large literature of this field in detail, especially since three relatively recent articles provide extensive reviews of it. The first, by R. H. Muller [109], gives abstracts of some 55 relevant publications prior to 1944, including most of those described in Section 3.1. (See Clayton [23] for criticism of Muller's abstracts of Clayton's work.) The second, by W. Bleeker [10], includes references to a number of continental articles not mentioned by Muller, and analyzes a number of proposed measures in detail, especially as regards the behavior of a predictor who knows that his predictions will be compared with actuality by a particular measure. The third, by G. W. Brier and R. A. Allen [17] discusses key publications appearing up to 1951. In the following paragraphs, we want to mention a few articles of particular interest to us, especially some published since the three surveys cited above.

The simplest case of interest to the meteorologists is where there is no order in the classifications and an asymmetrical interest in the two classifications. Sometimes the classifications are different, as when one is considering a particular qualitative variable as a predictor of qualitative weather. For this case, a measure of association based on the Shannon-Wiener information notion has been suggested by J. L. Holloway, Jr. and M. A. Woodbury [81] and has been used by several meteorologists, notably I. I. Gringorten and his colleagues. We have referred to it in Section 4.6. Gringorten [70, pp. 69–70] also suggests independently the same proportional prediction measure described in [66, Section 9]. This measure is very natural if we think of the possibility of making probabilistic, rather than categorical, forecasts, a possibility to which we shall recur in a few paragraphs. Gringorten's article also gives a brief general survey of measures of association in the meteorological context.

Sometimes the two classifications are the same, as when one is considering association between a categorical forecast and a categorical event, with both forecast and event classified in the same way. In this case of "forecast verification" both the above measures may be used, as well as others that take the identity of the two classifications explicitly into account. The use of association measures in connection with meteorological prediction, both with and without order taken into account, is considered by van der Bijl [140].

A more complex situation is that in which some third classification is brought into the picture. One important example is the three-way classification: forecast weather—observed weather—weather at time of forecast. Here interest is usually centered in the extent to which the forecaster can improve on persistence forecasting or on forecasting based on climatic information conditional upon weather at forecast time. Some materials referred to in the above paragraphs bear upon this situation; we should also like to cite two articles by Gringorten, [68] and [71a], and a closely related report by Gringorten, Lund, and Miller [69]. These references use scoring schemes with scores based on probabilities. Gringorten [68] makes it very clear that the appropriate measure depends upon the question being asked. In [71], Gringorten works on the sampling problem for measures based on scores.

An interesting problem is that of the construction of meaningful measures of association when the forecast is not categorical, but rather is itself a discrete probability distribution over a set of weather categories. Thus, for example, a prediction might be

No rain (probability .1)
Light rain (probability .6)
Heavy rain (probability .3)

and this prediction would be compared with that one of the three possibilities that later actually occurred. Suggestions for this kind of forecast prediction appear to go back at least to World War I, but it seems to have become of general interest only recently. Two recent articles relating to probabilistic forecasts are by G. W. Brier [16] and W. G. Leight [98].

If we attempt to construct a measure of association between probabilistic forecasts and the actual events later observed, we are faced with association between an essentially continuous distribution on a $k-1$ dimensional simplex (k categories, probabilities for each that sum to one) and a discrete distribution on k points (for the actual events).

Several articles take up Peirce's 1884 theme relating to economic losses as an important factor in evaluating forecast utility. For the 2×2 case, we refer to E. G. Bilham [9], H. C. Bijvoet and W. Bleeker [8], J. C. Thompson [135], J. C. Thompson and G. W. Brier [136], and G. W. Brier [18]. Gringorten [68 and 71a] considers more general cases by means of scores based directly on net losses.

4.11. *Association between species. Forbes, Cole, Goodall.* In the ecological literature there is a series of articles dealing with 2×2 cross classifications of the following kind:

NUMBERS OF AREAS IN WHICH SPECIES A AND SPECIES B ARE OR ARE NOT FOUND

		B: Found	B: Not Found	
A	Found	N_{11}	N_{12}	
	Not Found	N_{21}	N_{22}	
				n

Thus, for example, in N_{11} out of n marshes examined, grasses of species A and B are both found, while in N_{12} out of n marshes, species A is found but not species B.

A review and bibliography of ecological articles dealing with measures of association in this context is given by Goodall [64, pp. 221–3]. The series seems to have started with an article by Forbes [49] in 1907, followed by a long gap, and then a number of more recent articles. Of these, a particularly extensive one is by Cole [28], in which Benini's measure (see Section 3.3) was independently proposed.

4.12. *Association between anthropological traits. Tylor, Clements, Wallis, Driver, Kroeber, Chrétien, Kluckhohn, and others.* We have already discussed (Section 4.4) a proposal by the anthropologist, F. Boas. We now turn to a more special case than the one discussed by Boas, the 2×2 cross classification. Writers in the fields of anthropology and linguistics have long been concerned with 2×2 cross classifications similar to those discussed in the last section. The earliest paper of which we know that deals at all with measures of association in these fields is by Edward B. Tylor [139] in 1889. Tylor discussed many examples of association between cultural traits, some dichotomous and some trichotomous, but he contented himself with observing sizable apparent deviations from independence and did not suggest any numerical measures of association. In the ensuing discussion Francis Galton said [139, p. 270] that "... the degree of interdependence might with advantage be expressed in terms of a scale in which 0 represented perfect independence and 1 complete concurrence." We now list and discuss briefly those subsequent papers of which we know in this area that seem to us most germane to our survey.

In 1911, Jan Czekanowski [30], explicitly carrying Tylor's work forward, discussed the use of Yule's Q in ethnology and anthropology. Czekanowski also published a number of further papers dealing with 2×2 classifications.

In 1926, Forrest E. Clements and others [24] used the values of χ^2 and the resulting P-values in an examination of traits held in common by various Polynesian societies. An interesting controversy between Clements and Wilson D. Wallis [25, 143] followed. Wallis attacked Clements and his coauthors for using oversimplifying statistical methods and for drawing unjustified anthropological conclusions by these methods. Another article by Clements [26], discussing Q and ϕ prefixed by the appropriate $\pm$ sign, appeared in 1931. A quite recent article [27] by Clements goes over the same ground with added comments on subsequent literature.

In 1932, H. E. Driver and A. L. Kroeber [38] commented on the Clements-Wallis controversy, and used the following three measures in analyzing association between various pairs of societies:

$$\frac{\rho_{11}}{2}\left(\frac{1}{\rho_{1\cdot}}+\frac{1}{\rho_{\cdot 1}}\right), \qquad \frac{\rho_{11}}{\sqrt{\rho_{1\cdot}\rho_{\cdot 1}}}, \qquad \frac{\rho_{11}}{1-\rho_{22}}.$$

The 2×2 cross classifications to which these were applied referred to populations of traits, and took the following form:

		Society B: Has	Society B: Has not	
Society A	Has	ρ_{11}	ρ_{12}	$\rho_{1\cdot}$
	Has not	ρ_{21}	ρ_{22}	$\rho_{2\cdot}$
		$\rho_{\cdot 1}$	$\rho_{\cdot 2}$	

so that ρ_{12}, for example, is the proportion of traits observed present in Society A and absent in Society B.

In 1937, A. L. Kroeber and C. D. Chrétien [94] applied 2×2 measures of association to linguistic classification. Several measures were discussed and compared. Such application to linguistics continued in several articles, notably [20]. A recent article by Chrétien in this line is [22]. It is interesting to observe that the article immediately following [22], by Joseph H. Greenberg [67], is one of the few instances we know in which descriptive statistics are constructed so as to have operational interpretations in the sense that we have discussed. Greenberg's suggestions relate to measuring concentration in a single classification, or multinomial, population.

In 1939, a critical survey of the application of measures of association to ethnological data was published by Clyde Kluckhohn [90]. This very interesting article contains an extensive bibliography, and it marshals many arguments for and against the use of measures of association in anthropological contexts.

Driver [39], in the same year, compared in detail formal properties and relations between some eight 2×2 measures of association. He was much concerned with the effect of nonuniform marginal distributions on comparisons between values of 2×2 measures.

In 1945, Chrétien [21] discussed a number of basic points, including several analyzed by Kluckhohn, regarding the use of measures of association. Here, for almost the first and only time in this line of papers, we find the problem of interpretation raised as Chrétien says (p. 488): "Primary in importance, it seems to me, is the need to determine more precisely the meaning of the scale of association. All association studies to date have confined their attention to the high positive values."

Finally, we wish to cite a 1953 survey article by Driver [40]. In its section on ethnology and social anthropology, there appears a discussion of measures of association for the 2×2 case.

4.13. *Other suggestions.* We conclude by listing a few other suggestions relating to measures of association that do not fall naturally into the above classification.

Harris, Pearson. In a number of articles by J. A. Harris and others, [74], [75] and [76], there is a discussion of the following situation: Sometimes the existence of observations (individuals) in certain cells of a cross classification table is arithmetically, physically, or otherwise impossible. Harris and his co-authors discuss the effect of this inherent emptiness of some cells on certain traditional measures of association, and suggest modifications of these measures. K. Pearson, commenting on Harris' papers in [113], discusses the computation of the coefficient of mean square contingency when careful *a priori* consideration indicates that for certain cells the appearance of individuals in those cells is impossible. With the use of measures of association that have operational meaning, rather than the coefficient of mean square contingency, the occurrence of zero frequencies in certain cells does not seem to us to be of special significance. See Sec. 2.1. The *a priori* considerations leading to the belief about zero frequencies may, however, suggest alternative ways of setting up the classifications that are more meaningful.

Irwin. In 1934, J. O. Irwin [84] commented on measures of association and emphasized the importance of relating the use of such measures to the goals of the particular investigation at hand. He says (p. 87) that " . . . we should [not] do away with correlation coefficients or other measures of association, but should try to make the end point of our statistical analysis not a single coefficient which may be hard to interpret, but a result bearing a 'physical' meaning; the more easily the result may be understood by an intelligent layman, the better we should regard it as expressed." Irwin ends his article by describing a particular case of careful and useful analysis based on measures of association applied to the data in various ways.

It seems to us that, when the operational interpretation viewpoint towards association measures is taken, one is automatically influenced away from sterile arguments about which measure is "best." For if different measures reflect different aspects of the population, no one is best in any abstract sense (although one may be most appropriate in a given case) and there is no reason why more than one should not be used. An analogy is to ask about measures of size for human beings. One might suggest weight, height, volume, girth, etc., but no one of these is best except perhaps in a particular context.

Lakshmanamurti. In [97], M. Lakshmanamurti suggested a rather complex measure of association for the 2×2 case and compared it with Yule's Q.

Fairfield Smith. In a recent article [126a] H. Fairfield Smith has complained entertainingly about the difficulty of interpreting conventional measures of association. Most of his article shows by example how one may compare two sample cross classifications by forming simple chi-square tests that emphasize some specific aspect of possible difference between the cross classifications.

We end this paper with a quotation [126a, pp. 72–3] that expresses Smith's dismay about the vague or nonexistent meaning of most association measures.

> "What can be the use to know that ghosts in my lord's and lady's chambers each wore a sash with the symbol .6 if we do not know how the sash or its decoration may reflect the more earthy bodies from which the ghosts have been supposed to emanate?"

REFERENCES

[1] Andreoli, Giulio, "Sulla definizione di certi indici, relativi a caratteri di omogamia, in problemi statistici," *Rendiconto dell'Accademia delle Scienze Fisiche e Matematiche (Classe della Società Reale di Napoli)*, Ser. 4, 4 (1934), 36–41.

[2] Andreoli, Giulio, "Teoria generale di certi indici nei fenomeni statistici (omagamia, endogamia, diffusione, etc.)," *Rendiconto dell'Accademia delle Scienze Fisiche e Matematiche (Classe della Società Reale di Napoli)*, Ser. 4, 5 (1935), 108–24.

[3] Bachi, Roberto, "Gli indici della attrazione matrimoniale," *Giornale degli Economisti e Rivista di Statistica*, 69 (1929), 894–938.

[4] Benini, Rodolfo, *Principii di Demografia*, Firenzi, G. Barbèra, 1901. No. 29 of *Manuali Barbèra di Scienze Giuridiche Sociali e Politiche.*

[5] Benini, Rodolfo, "Grupi chiusi e gruppi aperti in alcuni fatti colletivi di combinazioni," *Bulletin de l'Institut International de Statistique*, 23 (1928), 362–83.

[6] Berkson, Joseph, "Cost utility as a measure of the efficiency of a test," *Journal of the American Statistical Association*, 42 (1947), 246–55.

[7] Berkson, Joseph, "The statistical study of association between smoking and lung cancer," *Proceedings of the Staff Meetings of the Mayo Clinic*, 30 (1955), 319–48.

[8] Bijvoet, H. C., and Bleeker, W., "The value of weather forecasts," *Weather*, 6 (1951), 36–9.

[9] Bilham, E. G., "A problem in economics," *Nature*, 109 (1922), 341–2.

[10] Bleeker, W., "The verification of weather forecasts," *Mededeelingen en Verhandelingen*, Ser. B, Deel 1, No. 2 (1946); Koninklijk Nederlandsch Meteorologisch Instituut, No. 125.

[11] Boas, Franz, "The measurement of differences between variable quantities," *Journal of the American Statistical Association*, 18 (1922–23), 425–45.

[12] Bogue, Donald J., *The Structure of the Metropolitan Community.* University of Michigan, Ann Arbor, 1949.

[12a] Bonferroni, C. E., "Nuovi indici di connessione fra variabili statistiche," *Pubblicazioni dell'Istituto di Statistica*, Università Commerciale "Luigi Bocconi," Milan. Vol. I (1942) of *Studi Sulla Correlazione e Sulla Connessione*, 57–100.

[13] Bonferroni, Carlo Emilio, "Indici unilaterali e bilaterali di connessione," 49–60 in: Mazzoni, Pacifico, and Lasorsa, Giovanni (ed.), *Studi in Memoria di Rodolfo Benini*, Facoltà di Economia e Commercio, Università degli Studi di Bari, Bari, 1956.

[14] Brambilla, Francesco, *La Variabilità Statistica a Due Dimensioni. La Teoria Statistica della Correlazione e della Connessione.* Università Commerciale "Luigi Bocconi," Istituto di Statistica, Casa Editrice Ambrosiana, Milan, 1949. This volume is part 2 of Brambilla's *Manuale di Statistica.*

[15] Bresciani, Costantino, "Sui metodi per la misura delle correlazioni," *Giornale degli Economisti*, 38 (1909), 401–14, 491–522.

[16] Brier, Glenn W., "Verification of forecasts in terms of probability," *Monthly Weather Review*, 78 (1950), 1–3.

[17] Brier, Glenn W., and Allen, Roger A., "Verification of weather forecasts," 841–8 in: Malone, Thomas F. (ed.), *Compendium of Meteorology*, American Meteorological Society, Boston, 1951.

[18] Brier, G. W., "The effect of errors in estimating probabilities on the usefulness of probability forecasts," *Bulletin of the American Meteorological Society*, 38 (1957), 76–8.

[19] Cartwright, Desmond S., "A rapid non-parametric estimate of multi-judge reliability," *Psychometrika*, 21 (1956), 17–29.

[20] Chrétien, C. Douglas, "The quantitative method for determining linguistic relationships. Interpretation of results and tests of significance," *University of California Publications in Linguistics*, 1 (1943–8), 11–19. Published in 1943.

[21] Chrétien, C. Douglas, "Culture element distributions: XXV. Reliability of statistical procedures and results," *University of California Publications in Anthropological Records*, 8 (1942–5), 469–90. Published in 1945.

[22] Chrétien, C. Douglas, "Word distributions in Southeastern Papua," *Language*, 32 (1956), 88–108.

[23] Clayton, H. H., "On verification of weather forecasts," *Bulletin of the American Meteorological Society*, 25 (1944), 368.

[24] Clements, Forrest E., Schenck, Sara M., and Brown, T. K., "A new objective method for showing special relationships," *American Anthropologist*, 28 (1926), 585–604.

[25] Clements, Forrest E., "Quantitative method in ethnography," *American Anthropologist*, 30 (1928), 295–310.

[26] Clements, Forrest, "Plains Indian tribal correlations with sun dance data," *American Anthropologist*, 33 (1931), 216–27.

[27] Clements, Forrest E., "Use of cluster analysis with anthropological data," *American Anthropologist*, 56 (1954), 180–99.

[28] Cole, La Mont C., "The measurement of interspecific association," *Ecology*, 30 (1949), 411–24.

[29] Cramér, Harald, "Remarks on correlation," *Skandinavisk Aktuarietidskrift*, 7 (1924), 220–40.

[30] Czekanowski, Jan, "Objective Kriterien in der Ethnologie," *Mitteilungen der Anthropologischen Gesellschaft in Wien*, 42 (1912), Sitzungsberichte (1911–12) Section, 17–21.

[31] de Meo, G., "Su di alcuni indici atti a misurare l'attrazione matrimoniale in classificazioni dicotome," *Rendiconto dell'Accademia delle Scienze Fisiche e Matematiche (Classe della Società Reale di Napoli)*, Ser. 4, 4 (1934), 62–77.

[32] Deuchler, Gustav, "Über die Methoden der Korrelationsrechnung in der Pädagogik und Psychologie," *Zeitschrift für Pädagogische Psychologie und Experimentelle Pädagogik*, 15 (1914), 114–31, 145–59, and 229–42.

[33] Deuchler, Gustav, "Über die Bestimmung von Rangkorrelationen aus Zeugnisnoten," *Zeitschrift für Angewandte Psychologie*, 12 (1916–17), 395–439.

[34] Deuchler, Gustav, "Über die Bestimmung einseitiger Abhängigkeit in pädagogisch-psychologischen Tatbeständen mit alternativer Variabilität," *Zeitschrift für Pädagogische Psychologie und Experimentelle Pädagogik*, 16 (1915), 550–66.

[35] Doolittle, M. H., "The verification of predictions" (abstract), *Bulletin of the Philosophical Society of Washington*, 7 (1885), 122–7. A summary of this article appears in *The American Meteorological Journal*, 2 (1885–6), 327–9.

[36] Doolittle, M. H., "Association ratios" (abstract), *Bulletin of the Philosophical Society of Washington*, 10 (1888), 83–7 and 94–6.

[37] Dornbusch, Sanford M., and Schmid, Calvin F., *A Primer of Social Statistics*, McGraw-Hill, New York, 1955.

[38] Driver, H. E., and Kroeber, A. L., "Quantitative expression of cultural relationships," *University of California Publications in American Archaeology and Ethnology*, 31 (1931–3), 211–56.

[39] Driver, Harold E., "Culture element distributions: X. Northwest California," *University of California Publications in Anthropological Records*, 1(6) (1939), 297–433.

[40] Driver, Harold E., "Statistics in anthropology," *American Anthropologist*, 55 (1953), 42–59.

[41] Duncan, Otis Dudley, Ohlin, Lloyd E., Reiss, Albert J., Jr., and Stanton, Howard R., "Formal devices for making selection decisions," *The American Journal of Sociology*, 58 (1953), 573–84.

[42] Duncan, Otis Dudley, and Duncan, Beverly, "A methodological analysis of segregation indexes," *American Sociological Review*, 20 (1955), 210–7.

[43] Eyraud, Henri, "Les principes de la mesure des corrélations," *Annales de l'Université de Lyon*, Ser. A (3), 1 (1936), 30–47.

[44] Faleschini, Luigi, "Indici di connessioni," *Contributi del Laboratorio di Statistica, Ser. 6, Università Cattolica del Sacro Cuore* [*Milan*], *Publicazioni, Nuova Serie*, 21 (1948), 112–51.

[45] Féron, Robert, "Mérites comparés des divers indices de corrélation," *Journal de la Société Statistique de Paris*, 88 (1947), 328–50. Discussion on pp. 350–2.

[46] Féron, Robert, "Information, Régression, Corrélation," *Publications de l'Institut de Statistique de l'Université de Paris*, 5(3–4) (1956), 113–215.

[47] Finley, Jno. P., "Tornado predictions," *The American Meteorological Journal*, 1 (1884), 85–8.

[48] Florence, P. Sargant, *Investment, Location, and Size of Plant*. Cambridge University Press, Cambridge, 1948.

[49] Forbes, S. A., "On the local distribution of certain Illinois fishes: an essay in statistical ecology," *Bulletin of the Illinois State Laboratory of Natural History*, 7 (1904–9), 273–303, plus 32 plates.

[50] Fréchet, Maurice, "A general method of constructing correlation indices," *Proceedings of the Mathematical and Physical Society of Egypt*, 3 (2) (1946), 13–20. (Additional note 3 (4) (1948), 73–4.)

[51] Fréchet, Maurice, "Anciens et nouveaux indices de corrélation. Leur application au calcul des retards économiques," *Econometrica*, 15 (1947), 1–30. (Errata pp. 374–5.)

[52] Fréchet, Maurice, "Sur les tableaux de corrélation dont les marges sont données," *Annales de l'Université de Lyon*, Sec. A (3), 14 (1951), 53–77.

[53] Fréchet, Maurice, "Sur les tableaux de corrélation dont les marges sont données," *Comptes Rendus Hebdomadaires des Scéances de l'Académie des Sciences, Paris*, 242 (1956), 2426–8.

[54] Galvani, L., "A propos de la communication de M. Thionet: L'école moderne de statisticiens italiens," *Journal de la Société de Statistique de Paris*, 88 (1947), 196–203.

[55] Gilbert, G. K., "Finley's tornado predictions," *The American Meteorological Journal*, 1 (1884), 166–72.

[56] Gini, Corrado, "Di una misura della dissomiglianza tra due gruppi di quantità e delle sue applicazioni allo studio delle relazioni statistiche," *Atti del Reale Istituto Veneto di Scienze, Lettere ed Arti*, Series 8, 74 (2) (1914–5), 185–213.

[57] Gini, Corrado, "Indice di omofilia e di rassomiglianza e loro relazioni col coefficiente di correlazione e con gli indici di attrazione," *Atti del Reale Istituto Veneto di Scienze, Lettere ed Arti*, Series 8, 74 (2), (1914–5), 583–610.

[58] Gini, Corrado, "Nuovi contributi alla teoria delle relazioni statistiche," *Atti del Reale Istituto Veneto di Scienze, Lettere ed Arti*, Series 8, 74 (2), (1914–5), 1903–42.

[59] Gini, Corrado, "Sul criterio di concordanza tra due caratteri," *Atti del Reale Istituto Veneto di Scienze, Lettere ed Arti*, Series 8, 75 (2), (1915–6), 309–31.

[60] Gini, Corrado, "Indici di concordanza," *Atti del Reale Istituto Veneto di Scienze, Lettere ed Arti*. Series 8, 75 (2), (1915–6), 1419–61.

[61] Gini, Corrado, "The contributions of Italy to modern statistical methods," *Journal of the Royal Statistical Society*, 89 (1926), 703–24.

[61a] Gini, Corrado, "Die Messung der Ungleichheit zweier Verteilungen, angewendet auf die Untersuchung von qualitativen Rassenmerkmalen," *Archiv für Mathematische Wirtschafts- und Sozialforschung*, 3 (1937), 167–84, plus two appendixes by Vittorio Castellano.

[62] Gini, Corrado, *Metodologia Statistica: la Misura dei Fenomeni Collettivi*, Vol. III, Part 2a, Chap. 55, 245–321, *Enciclopedia delle Matematiche Elementari*, edited by Luigi Berzolari, Ulrico Hoepli, Milan, 1948.

[63] Gini, Corrado, *Corso di Statistica*, a cura di S. Gatti e Benedetti, nuova edizione aggiornata, Università degli Studi di Roma, Facoltà di Scienze Statistiche, Demografiche ed Attuariali, Libreria Eredi Virgilio Veschi, Rome, 1954–5.

[63a] Glass, D. V. (ed.), *Social Mobility in Britain*, Routledge and Kegan Paul, London, 1954.

[64] Goodall, D. W., "Quantitative aspects of plant distribution," *Biological Reviews of the Cambridge Philosophical Society*, 27 (1952), 194–245.

[65] Goodman, Leo A., "On urbanization indices," *Social Forces*, 31 (1953), 360–2.

[66] Goodman, Leo A., and Kruskal, William H., "Measures of association for cross classifications," *Journal of the American Statistical Association*, 49 (1954), 723–64.

[67] Greenberg, Joseph H., "The measurement of linguistic diversity," *Language*, 32 (1956), 109–15.

[68] Gringorten, Irving, I., "The verification and scoring of weather forecasts," *Journal of the American Statistical Association*, 46 (1951), 279–96.

[69] Gringorten, Irving I., Lund, Iver A., and Miller, Martin A., "A program to test skill in terminal forecasting," Air Force Surveys in Geophysics, No. 80, Geophysics Research and Development Command, June, 1955.

[70] Gringorten, Irving I., "Methods of objective weather forecasting," pp. 57–92 in: Landsberg, H. E. (ed.), *Advances in Geophysics*, Vol. 2, Academic Press, New York, 1955.

[71] Gringorten, Irving I., "Tests of significance in a verification program," *Journal of Meteorology*, 12 (1955), 179–95.

[71a] Gringorten, Irving I., "On the comparison of one or more sets of probability forecasts," *Journal of Meteorology*, 15 (1958), 283–7.

[72] Gringorten, Irving I., Lund, Iver A., and Miller, Martin A., "The construction and use of forecast registers," Geophysical Research Papers, No. 53, Geophysics Research Directorate, Air Force Cambridge Research Center, Air Research and Development Command, June, 1956.

[73] Guilford, J. P., *Fundamental Statistics in Psychology and Education*, McGraw-Hill, New York, First edition, 1942, second edition, 1950.

[73a] Halphen, Étienne, "L'analyse intrinsèque des distributions de probabilité," *Publications de l'Institut de Statistique de l'Université de Paris*, 6 (2) (1957), 79–159.

[74] Harris, J. Arthur, and Treloar, Alan E., "On a limitation in the applicability of the contingency coefficient," *Journal of the American Statistical Association*, 22 (1927), 460–72.

[75] Harris, J. Arthur, Treloar, Alan E., and Wilder, Marian, "On the theory of contingency II. Professor Pearson's note on our paper on contingency," *Journal of the American Statistical Association*, 25 (1930), 323–7.

[76] Harris, J. Arthur, and Tu, Chi, "A second category of limitation in the applicability of the contingency coefficient," *Journal of the American Statistical Association*, 24 (1929), 367–75.

[77] Hazen, H. Allen, "Verification of tornado predictions," *American Journal of Science*, Ser. 3, 34 (1887), 127–31.

[78] Höffding, Wassilij, "Massstabinvariante Korrelationstheorie," *Schriften des Mathematischen Instituts und des Instituts für Angewandte Mathematik der Universität Berlin*, 5 (3) (1940), 181–233.

[79] Höffding, Wassilij, "Massstabinvariante Korrelationsmasse für diskontinuienliche Verteilungen," *Archiv für Mathematische Wirtschafts- und Sozialforschung*, 7 (1941), 49–70.

[80] Höffding, Wassilij, "Stochastische Abhängigkeit und funktionaler Zusammenhang," *Skandinavisk Aktuarietidskrift*, 25 (1942), 200–27.

[81] Holloway, J. Leith, Jr., and Woodbury, Max A., "Application of information theory and discriminant function analysis to weather forecasting and forecast verification," Technical Report No. 1, Meteorological Statistics Project, Institute for Cooperative Research, University of Pennsylvania, Feb., 1955.

[82] Hoover, Edgar M., Jr., "The measurement of industrial localization," *Review of Economic Statistics*, 18 (1936), 162–71.

[83] Hoover, Edgar M., Jr., "Interstate redistribution of population, 1850–1940," *Journal of Economic History*, 1 (1941), 199–205.

[84] Irwin, J. O., "Correlation methods in psychology," *British Journal of Psychology, General Section*, 25 (1934–5), 86–91.

[84a] Johnson, H. M., "Maximal selectivity, correctivity and correlation obtainable in a 2×2 contingency table," *American Journal of Psychology*, 58 (1945), 65–8.

[85] Jordan, Charles, "Les coefficients d'intensité relative de Körösy," *Revue de la Société Hongroise de Statistique* (*Magyar Statisztikai Társaság, Revue*), 5 (1927), 332–45.

[86] Jordan, Charles, "Critique de la corrélation au point de vue des probabilités," *Actes du Colloque Consacré à la Théorie des Probabilités*, No. 740 of *Actualités Scientifiques et Industrielles*, Hermann, Paris, 1938.

[87] Jordan Károly, "A Korreláció számitása (Sur le calcul de la corrélation) I," *Magyar Statisztikai Szemle Kiadványai* (Központi Statisztikai Hivatal), (1941) 1 Szám. Hungarian with French summary.

[87a] Katz, Elihu, and Lazarsfeld, Paul E., *Personal Influence*, The Free Press, Glencoe, Illinois, 1955.

[88] Kendall, Patricia, *Conflict and Mood*, The Free Press, Glencoe, Illinois, 1954.

[89] Klein, Herman J., "Ergebnisse rationeller Prüfungen von Wetterprognosen und deren Bedeutung für die Praxis," *Wochenschrift für Astronomie, Meteorologie und Geographie*, NF 28 (1885), 9–14, 20–2, 25–8, 35–7, 41–4, and 57–61.

[90] Kluckhohn, Clyde, "On certain recent applications of association coefficients to ethnological data," *American Anthropologist*, 41 (1939), 345–77.

[91] Köppen, Wladimir, "Die Aufeinanderfolge der unperiodischen Witterungserscheinungen nach den Grundsätzen der Wahrscheinlichkeitsrechnung," *Repertorium für Meteorologie*, Akademiia Nauk, Petrograd, 2 (1870–1), 189–238.

[92] [Köppen, W.?], "Eine rationelle Methode zur Prüfung der Wetterprognosen," *Meteorologische Zeitschrift*, 1 (1884), 397–404.

[93] Körösi, Joseph, "Kritik der Vaccinations-Statistik und neue Beiträge zur Frage des Impfschutzes," *Transactions of the Ninth International Medical Congress*, Washington, D. C., 1887, Vol. 1, 238–418.

[94] Kroeber, A. L., and Chrétien, C. D., "Quantitative classification of Indo-European language," *Language* 13 (1937), 83–103.

[95] Kruskal, William H., "Historical note on the Wilcoxon unpaired two-sample test," *Journal of the American Statistical Association*, 52 (1957), 356–60.

[96] Kruskal, William H., "Ordinal measures of association," *Journal of the American Statistical Association* 53 (1958), pp. 814-61.

[97] Lakshmanamurti, M., "Coefficient of association between two attributes in statistics," *Proceedings of the Indian Academy of Sciences*, Ser. A, 22 (1945), 123–33.

[98] Leight, Walter G., "The use of probability statements in extended forecasting," *Monthly Weather Review*, 81 (1953), 349–56.

[99] Linfoot, E. H., "An informational measure of association," *Information and Control*, 1 (1957), 85–9.

[100] Lipps, Gottl. Friedr., "Die Bestimmung der Abhängigkeit zwischen den Merkmalen eines Gegenstandes," *Berichte über die Vorhandlungen der Königlich Sächsischen Gesellschaft der Wissenschaften zu Leipzig, Mathematisch-Physische Klasse*, 57 (1905), 1–32.

[101] Lipps, G. F., *Die Psychischen Massmethoden*, F. Vieweg und Sohn, Braunschweig, 1906. Vol. 10 of the monograph series, *Die Wissenschaft.*

[102] Loevinger, Jane, "A systematic approach to the construction and evaluation of tests of ability," *Psychological Monographs*, 61 (4), (1947).

[103] Loevinger, Jane, "The technic of homogeneous tests compared with some aspects of 'scale analysis' and factor analysis," *Psychological Bulletin*, 45 (1948), 507–29.

[104] Long, John A., "Improved overlapping methods for determining the validities of test items," *Journal of Experimental Education*, 2 (1934), 264–8.

[105] Lorey, W., "Neue Abhängigkeitsmasse für zufällige Grossen," *Deutsches Statistisches Zentralblatt*, 25 (1933), 153–4.

[106] Mainland, Donald, "The risk of fallacious conclusions from autopsy data on the incidence of diseases with applications to heart disease," *American Heart Journal*, 45 (1953), 644–54.

[107] McGill, William J., "Multivariate information transmission," *Psychometrika*, 19 (1954), 97–116.

[108] Meyer, Hugo, *Anleitung zur Bearbeitung Meteorologischer Beobachtungen für die Klimatologie*, Julius Springer, Berlin, 1891.

[109] Muller, Robert Hans, "Verification of short-range weather forecasts (a survey of the literature)," *Bulletin of the American Meteorological Society*, 25 (1944), 18–27, 47–53, and 88–95.

[110] Niceforo, Alfredo, *La Misura della Vita*, Fratelli, Bocca, Torino, 1919.

[111] Niceforo, Alfredo, *La Méthode Statistique et ses Applications aux Sciences Naturelles aux Sciences Sociales et à l'Art*, Marcel Giard, Paris, 1925. Translation by R. Jacquemin, with some changes, of Niceforo's *Il Metodo Statistico* (1923).

[112] Pearson, Karl, "On a coefficient of class heterogeneity or divergence," *Biometrika*, 5 (1906), 198–203.

[113] Pearson, Karl, "On the theory of contingency. I. Note on Professor J. Arthur Harris' paper on the limitation in the applicability of the contingency coefficient," *Journal of the American Statistical Association*, 25 (1930), 320–3.

[114] [Pearson, Karl?], "Remarks on Professor Steffensen's measure of contingency. Editorial," *Biometrika*, 26 (1934), 255–60.

[115] Peirce, C. S., "The numerical measure of the success of predictions" (letter to the editor), *Science*, 4 (1884), 453–4.

[116] Pietra, G., "The theory of statistical relations with special reference to cyclical series," *Metron*, 4 (3–4), (1924–5), 383–557.

[117] Pollaczek-Geiringer, H., "Bemerkungen zur Korrelations-theorie," *Verhandlungen des Internationalen Mathematiker-Congress*, Zurich (1932), 2, 229–30.

[118] Pollaczsk-Geiringer, H., "Korrelationmessung auf Grund der Summenfunktion," *Zeitschrift für Angewandte Mathematik und Mechanik*, 13 (1933), 121–4.

[119] Quetelet, M. A., *Letters addressed to H.R.H. the Grand Duke of Saxe Coburg and Gotha on the Theory of Probabilities as Applied to the Moral and Political Sciences* (translated from the French by Olinthus Gregory Downs), Charles and Edwin Layton, London, 1849.

[120] Reuning, H., "Evaluation of square contingency tables. A simple method of condensation for small samples," *Bulletin, National Institute for Personnel Research*, (South African Council for Scientific and Industrial Research, Johannesburg), 4 (1952), 160–7.

[121] Saile, Tivadar Antal, *Influence de Joseph de Körösy Sur l'Évolution de la Statistique*, Kiadja a Magyar Tudományos Akadémia, Budapest, 1927. Hungarian with French summary.

[122] Salvemini, Tommaso, "Su alcuni indici usati per la misura delle relazioni statistiche," *Statistica*, 7 (1947), 200–18.

[123] Salvemini, T., "Nuovi procedimenti di calcolo degli indici di dissomigliaza e di connessione," *Statistica*, 9 (1949), 3–27.

[124] Salvemini, Tommaso, "Su alcuni aspetti della dissomiglianza e della concordanza con applicazione alle distribuzioni degli sposi secondo l'età," *Bulletin de l'Institut International de Statistique*, 34 (2), (1954), 283–300.

[125] Salvemini, Tommaso, "Fondamenti razionali, schemi teorici e moderni aspetti delle relazioni tra due o più variabili," *Atti della XV Riunione Scientificà, Societa Italiana di Statistica*, 1955.

[126] Savorgnan, Franco, "La misura dell'endogamia e dell'omogamia," in *Proceedings of the International Congress for Studies on Population (Rome, 7th–10th September 1931–IX)*, Vol. X, 9–37. Published by the Comitato Italiano per lo Studio dei Problemi della Popolazione, through the Istituto Poligrafico dello Stato, Rome, 1934.

[126a] Smith, H. Fairfield, "On comparing contingency tables," *The Philippine Statistician*, 6 (1957), 71–81.

[127] Steffensen, J. F., "Deux problèmes du calcul des probabilités," *Annales de l'Institut Henri Poincaré*, 3 (1932–3), 319–44. Plus one non-numbered page of errata.

[128] Steffensen, J. F., "On certain measures of dependence between statistical variables," *Biometrika*, 26 (1934), 250–5.

[129] Steffensen, J. F., "On the ω test of dependence between statistical variables," *Skandinavisk Aktuarietidskrift*, 24 (1941), 13–33.

[130] Striefler, Heinrich, "Zur Methode der Rangkorrelation nach Tönnies," *Deutsches Statistisches Zentralblatt*, 23 (1931), 128–35 and 160–7.

[131] "Student" [W. S. Gosset], "An experimental determination of the probable error of Dr Spearman's correlation coefficients," *Biometrika*, 13 (1920–1), 263–82. Reprinted in: Pearson, E. S., and Wishart, John (ed.), *"Student's" Collected Papers*, Biometrika Office, University College, London, 1942. Pp. 70–89.

[132] Thionet, Pierre, "L'école moderne de statisticiens italiens," *Journal de la Société de Statistique de Paris*, 86 (1945), 245–55, and 87 (1946), 16–34.

[133] Thionet, Pierre, "Résponse de l'auteur de la communication du 21 novembre 1945," *Journal de la Société de Statistique de Paris*, 88 (1947), 203–8.

[134] Thirring, Gustave, obituary of Joseph de Körösy, *Bulletin de l'Institut International de Statistique*, 16 (1907–8), 150–4.

[135] Thompson, J. C., "On the operational deficiencies in categorical weather forecasts," *Bulletin of the American Meteorological Society*, 33 (1952), 223–6.

[136] Thompson, J. C., and Brier, G. W., "The economic utility of weather forecasts," *Monthly Weather Review*, 83 (1955), 249–54.

[137] Tönnies, Ferdinand, "Eine neue Methode der Vergleichung Statistischer Reihen (im Anschluss an Mitteilungen über kriminalstatistische Forschungen)," *Jahrbuch für Gesetzgebung, Verwaltung und Volkswirtschaft im Deutschen Reich*, 33 (1909), 699–720.

[138] Tönnies, Ferdinand, "Korrelation der Parteien in Statistik der Kieler Reichstagswahlen," *Jahrbücher für Nationalökonomie und Statistik*, 122 (1924), 663–72.

[139] Tylor, Edward B., "On a method of investigating the development of institutions; applied to laws of marriage and descent," *Journal of the Anthropological Institute of Great Britain and Ireland*, 18 (1888–9), 245–69. Discussion, 270–2.

[140] van der Bijl, Willem, "Statistische Bemerkungen zu H. Müller-Annen: 'Versuche von Langfristvorhersagen mit einer Kontingenzmethode' (Ann. Met. 6, S. 257) nebst weiteren Betrachtungen über den prognostischen Wert von Kontingenztabellen," *Annalen der Meteorologie*, 7 (1955–6), 289–302.

[141] Wahl, Eberhard W., "Das statistisch Entropieverhältnis, ein Hilfsmittel zur Lösung von Vorhersageproblemen," *Meteorologische Rundschau*, 8 (1955), 51–4.

[142] Wallis, W. Allen, and Roberts, Harry V., *Statistics, A New Approach*, The Free Press, Glencoe, Illinois, 1956.

[143] Wallis, Wilson D., "Probability and the diffusion of culture traits," *American Anthropologist*, 30 (1928), 94–106.

[144] Weida, Frank M., "On various conceptions of correlation," *Annals of Mathematics*, 29 (1928), 276–312.

[145] Williams, Josephine J., "Another commentary on so-called segregation indices," *American Sociological Review*, 13 (1948), 298–303.

[146] Wilson, Edwin B., "Morbidity and the association of morbid conditions," *The Journal of Preventive Medicine*, 4 (1930), 27–38.

[147] Wirth, Wilhelm, *Spezielle Psychophysische Massmethoden*, Urban and Schwarzenberg, Berlin and Vienna, 1920. Part of Abderhalden, Emil (ed.), *Handbuch der Biologischen Arbeitsmethoden;* Lieferung 4, Abt. 6, Methoden der Experimentellen Psychologie, A. Heft 1.

[148] Wood, Karl D., "Rapid correlation by an empirical method," *Journal of Educational Psychology*, 19 (1928), 243–51.

[149] Yule, G. Udny, "On the association of attributes in statistics: with illustrations from the material from the Childhood Society, &c," *Philosophical Transactions of the Royal Society of London, Ser. A*, 194 (1900), 257–319.

[150] Yule, G. Udny, "On the methods of measuring association between two attributes," *Journal of the Royal Statistical Society*, 75 (1912), 579–642. Discussion 643–52.

Reprinted from the JOURNAL OF THE AMERICAN STATISTICAL ASSOCIATION
June, 1963, Vol. 58, pp. 310-364

MEASURES OF ASSOCIATION FOR CROSS CLASSIFICATIONS III: APPROXIMATE SAMPLING THEORY*

LEO A. GOODMAN AND WILLIAM H. KRUSKAL
University of Chicago

The population measures of association for cross classifications, discussed in the authors' prior publications, have sample analogues that are approximately normally distributed for large samples. (Some qualifications and restrictions are necessary.) These large sample normal distributions with their associated standard errors, are derived for various measures of association and various methods of sampling. It is explained how the large sample normality may be used to test hypotheses about the measures and about differences between them, and to construct corresponding confidence intervals. Numerical results are given about the adequacy of the large sample normal approximations. In order to facilitate extension of the large sample results to other measures of association, and to other modes of sampling, than those treated here, the basic manipulative tools of large sample theory are explained and illustrated.

CONTENTS

	Page
1. INTRODUCTION AND SUMMARY	311
2. NOTATION AND PRELIMINARIES	313
3. MULTINOMINAL SAMPLING OVER THE WHOLE DOUBLE POLYTOMY	315
3.1. The Index λ_b	315
3.2. Use of Asymptotic Unit-normality	316
3.3. The Index λ_a	321
3.4. The Index λ	321
3.5. The Index γ	322
3.6. Measures of Reliability	330
3.7. Partial and Multiple Association	332
3.8. Sampling Experiments	334
4. MULTINOMIAL SAMPLING WITHIN EACH ROW (COLUMN) OF THE DOUBLE POLYTOMY	348
4.1. Preliminaries	348
4.2. The Index λ_b, with Marginal Row Probabilities Known	349
4.3. The Index λ_b^*	352
4.4. The Index τ_b, with Marginal Row Probabilities Known	353
5. FURTHER REMARKS	354
6. REFERENCES	355

APPENDIX

A1. Introduction	356
A2. A Basic Convergence Theorem	356

* This research was supported in part by the Army Research Office, the Office of Naval Research, and the Air Force Office of Scientific Research by Contract No. Nonr-2121(23); and in part by Research Grant NSF-G21058 from the Division of Mathematical, Physical, and Engineering Sciences of the National Science Foundation.

Part of Mr. Goodman's work on this paper was done at the Statistical Laboratory of the University of Cambridge under a Fulbright Award and a Social Science Research Council Fellowship. Part of Mr. Kruskal's work on this paper was done at the Department of Statistics, University of California, Berkeley.

We thank the following persons for helpful comments: H. E. Daniels, B. L. Fox, R. Kozelka, J. Mincer, I. R. Savage, R. Sneyers, and A. Stuart. We also wish to acknowledge computational work on this paper by M. DeGroot, W. Goldfarb, and G. Pinkham.

A3. The Delta Method Theorem 356
A4. Sample Maxima 357
A5. Asymptotic Behavior of L_b 358
A6. Asymptotic Behavior of L 360
A7. Asymptotic Behavior of G 361

1. INTRODUCTION AND SUMMARY

PROBLEMS connected with measuring the degree of association between two or more cross classifications or polytomies were considered by us in [8]. A number of possible measures or indexes of association were discussed for situations in which the meaning of the term "degree of association" is not completely clear-cut within some precisely stated model. The central theme of [8] was that measures of association should have operationally meaningful interpretations that are relevant in the contexts of empirical investigations in which the measures are used.[1]

A supplementary discussion [9] presented further measures of association, together with historical and bibliographical material.

The discussion in references [8] and [9] supposed almost throughout that the parent population is known, so that no sampling problems arise. In the present paper, we develop, by asymptotic (i.e., large sample) methods, approximate sampling theory for the measures considered in [8]; without having such a theory in practical form, the measures of [8] are of limited use. We use this sampling theory in connection with testing hypotheses about the measures and establishing confidence intervals for the measures. We also include here some material about the adequacy of the asymptotic approximations; we expect to present further material in a later publication.

The notation of [8] will be used freely, but we shall generally try to recapitulate for the reader's convenience. We restrict ourselves to sample sizes fixed in advance, and most, but not all, of this paper deals with the case of two polytomies or cross classifications, A with α classes, and B with β classes. The population or true frequency for the cell with A classification A_a and B classification B_b is denoted by ρ_{ab}; we set $\rho_{a\cdot} = \sum_b \rho_{ab}$ and $\rho_{\cdot b} = \sum_a \rho_{ab}$.

In developing an asymptotic sampling theory for the measures proposed in [8], a number of considerations arise.

i. Sampling methods. There are several possible sampling methods. For example, one may choose a random sample, in either the sense of "with replacement" (infinite population) or "without replacement," from a population of individuals that is cross classified into the $\alpha\beta$ cells obtained by crossing classifications A and B. In this case, the sampling method leads to a multinomial (with replacement) or to a generalized hypergeometric (without replacement) distribution.

---

[1] The measures of association considered in our papers may be appropriate in situations where little or no structural information is available about the true relative frequencies in the cells of the cross classification. If, contrariwise, a structural parametric model is assumed, it will often be the case that one or more of its parameters will have obvious meanings as measures of association within the terms of the assumed model.

We do not think it meaningful to speak of *the* most appropriate measure of association for most of the unstructured situations we have in mind. What is central is that the association measures used should have meaningful interpretations; it is quite likely that several measures, each with its own interpretation, might all be useful in a given situation.

Alternatively, one might sample independently *within* each A_a class or row of the $A \times B$ cross classification, obtaining α independent multinomial or generalized hypergeometric distributions. Then the relative sample sizes for the α classes of classification A would be of importance. Similarly, one might sample independently within each B_b class or column. There are still other possibilities, but we shall not deal with them here.

For the sake of simplicity, we assume infinite populations throughout; that is, sampling with replacement in the technical sense. Much of the work refers to the case of a multinomial sample over all the $\alpha\beta$ cells, but we do consider some cases of independent sampling in the rows or columns.

ii. Auxiliary knowledge. Auxiliary knowledge may vary considerably. Thus one may know marginal totals, the $\rho_{a\cdot}$'s, and/or the $\rho_{\cdot b}$'s. (For treatments of this, see [5], [25], and [6].) Or one may know the values of the b subscripts maximizing ρ_{ab} without actually knowing the numerical values of the ρ's. We generally assume no auxiliary knowledge, except for uniqueness assumptions to be described, but in Section 4 we do consider some cases in which auxiliary knowledge is utilized.

iii. Choice of estimators. The question of what estimator to use for a given measure is resolved here by using the obvious sample analogue of the population measure. In all cases but one, (4.2.1), this is the maximum likelihood estimator.[2] For testing and confidence intervals we again have used intuitively straightforward procedures based on the point estimators and their distributions.

iv. Uniqueness of maxima. Many of the asymptotic results depend on assumptions like this: there is just one value of b maximizing $\rho_{\cdot b}$; we shall generically say that such assumptions are those of uniqueness of maxima. Even when there is a unique value of b maximizing $\rho_{\cdot b}$, there may be other values of b for which $\rho_{\cdot b}$ is very near the maximum. In such cases, particularly large sample sizes may be needed before the asymptotic distributions we discuss become good approximations to the actual distributions.

v. Asymptotic approximations and their possible modifications. For each estimator and sampling method, our procedure is to find a function of the sample, an approximate standard error (ASE), such that the difference between the estimator and the true value of the measure being estimated, divided by the ASE, is for large samples approximately unit-normal (normal with zero mean and unit standard deviation);

$$(\text{estimator} - \text{true value})/(\text{ASE}) \approx N(0, 1)$$

for large samples.[3] Given such an approximation (corresponding to convergence in distribution), one may modify the estimator, the ASE, or both in many ways

[2] This paper studies multi-parameter situations, and, in fact, we generally consider $\alpha\beta-1$ independent parameters, one for the probability of each cell of the cross classification, modified by the restriction that these probabilities sum to one. We adopt the usual convention that, if $(\hat{\theta}_1, \cdots, \hat{\theta}_k)$ is the maximum likelihood estimator of $(\theta_1, \cdots, \theta_k)$, then by "the maximum likelihood estimator of $f(\theta_1, \cdots, \theta_k)$" we mean $f(\hat{\theta}_1, \cdots, \hat{\theta}_k)$. This convention is justified by the invariance of maximum likelihood estimation under reparameterization.

[3] Other approximations than normal ones are in principle possible.

without destroying mathematical convergence to unit-normality, but with possible improvement in the normal approximation for finite samples. It is also possible to consider a transformation of the true value and the estimator. Such modifications or transformations have been widely used to improve asymptotic approximations. (For example, corrections for continuity may be viewed from this standpoint.) We shall present a few possible transformations with their corresponding ASE's, and we expect to present further material in a later publication.

2. NOTATION AND PRELIMINARIES

A sample of n individuals is drawn in some specified manner. The A and B classifications of each member of the sample are observed. Let N_{ab} be the number of sample individuals that fall in the (A_a, B_b) cell; that is, N_{ab} is the number of individuals having A classification A_a and B classification B_b. Thus

$$\sum_{a=1}^{\alpha} \sum_{b=1}^{\beta} N_{ab} = n.$$

In the case of nonrestricted sampling, multinomial over the entire $\alpha \times \beta$ cross classification table, the marginals will be denoted in the conventional manner,

$$N_{a\cdot} = \sum_b N_{ab}, \qquad N_{\cdot b} = \sum_a N_{ab}. \tag{2.1}$$

In most other sampling methods, at least one set of these marginals will not be random but fixed in advance. In general, we shall use capital Latin letters for random variables and lower case Latin letters for corresponding fixed numbers, e.g., $n_{a\cdot} = \sum_b N_{ab}$ for fixed row marginals. The Latin letters to be used will, whenever feasible, be related to the Greek letters used for corresponding population quantities.

Thus R_{ab} (corresponding to ρ_{ab}) will be used for the proportion, N_{ab}/n, of observations in the (A_a, B_b) cell, $R_{a\cdot}$ for $N_{a\cdot}/n = \sum_b R_{ab}$ (when the row marginals are random), and so on; $\sum \sum_{ab} R_{ab} = 1$. It is convenient to work with the R_{ab}'s for present purposes, but the more important formulas will also be given in terms of the N_{ab}'s for convenience in applications. Special mention must be made of the notation for maxima. We denote by N_{am} the maximum over b of N_{ab}, with analogous notation for other maxima as follows:

$$\begin{aligned} N_{am} &= \operatorname*{Max}_b N_{ab} \\ N_{mb} &= \operatorname*{Max}_a N_{ab} \\ N_{m\cdot} &= \operatorname*{Max}_a N_{a\cdot} \quad (\text{or } n_{m\cdot} = \operatorname*{Max}_a n_{a\cdot}) \\ N_{\cdot m} &= \operatorname*{Max}_b N_{\cdot b} \quad (\text{or } n_{\cdot m} = \operatorname*{Max}_b n_{\cdot b}). \end{aligned} \tag{2.2}$$

The notation for R_{am}, $R_{\cdot m}$, etc. will be similar. It should be noted that a symbol like "$N_{m\cdot}$" is a single unit, meaning "the largest $N_{a\cdot}$".

From now until further notice (in Section 4), we assume multinomial sampling over all the $\alpha\beta$ cells. We shall throughout be considering asymptotic behavior as $n \to \infty$, so that in principle "n" should be attached to all symbols for random variables to indicate that we have in mind a sequence of samples: $N_{ab}^{(n)}$. For the sake of simplicity, however, we shall omit the "n."

One fact is basic: the $\alpha\beta$ random variables $\sqrt{n}(R_{ab}-\rho_{ab})$ jointly converge in distribution[4] to the multivariate normal distribution with all means zero, and with variances and covariances given by

$$\begin{aligned} &\operatorname{Var}[\sqrt{n}(R_{ab} - \rho_{ab})] = \rho_{ab}(1 - \rho_{ab}) \\ &\operatorname{Cov}[\sqrt{n}(R_{ab} - \rho_{ab}), \sqrt{n}(R_{a'b'} - \rho_{a'b'})] = -\rho_{ab}\rho_{a'b'} \quad (a \neq a' \text{ or } b \neq b'). \end{aligned} \tag{2.3}$$

(These means, variances, and covariances also are correct for any finite n.) A reference to this basic fact is p. 419 of [2]. It follows from the above that each R_{ab} converges in probability[4] to ρ_{ab}.

We now make, until further notice, the following assumptions about the population:

$$\begin{aligned} &\text{For each } a,\ \rho_{am} = \rho_{ab} \text{ for a unique value of } b. \\ &\text{For each } b,\ \rho_{mb} = \rho_{ab} \text{ for a unique value of } a. \\ &\rho_{\cdot m} = \rho_{\cdot b} \text{ for a unique value of } b. \\ &\rho_{m\cdot} = \rho_{a\cdot} \text{ for a unique value of } a. \end{aligned} \tag{2.4}$$

These are the assumptions mentioned in the preceding section. Without them no useful asymptotic theory for the distribution of estimators of the λ coefficients seems possible at the present time. (We need not always make all four assumptions; for example, in the case of λ_b the assumptions relating to ρ_{am} and $\rho_{\cdot m}$ will suffice. None of these uniqueness assumptions are needed in the case of γ.) An assumption that will be made throughout is that the population or true value of the measure of association in question is well defined.

Under these assumptions, and by convergence in probability, the probability that $R_{am} = R_{ab}$ approaches unity for that value of b such that $\rho_{am} = \rho_{ab}$. Similarly, the probability that $R_{\cdot m} = R_{\cdot b}$ approaches unity for that value of b such that $\rho_{\cdot m} = \rho_{\cdot b}$. Hence, for our asymptotic purposes, we may act as if R_{am} is taken on at that value of b such that $\rho_{am} = \rho_{ab}$ (see Section A4). Similarly, we may act as if $R_{\cdot m}$ is taken on at that value of b such that $\rho_{\cdot m} = \rho_{\cdot b}$. The same statements hold of course for R_{mb} and $R_{\cdot m}$.[5]

---

[4] This *convergence in distribution* means that the probability that the $\sqrt{n}\ (R_{ab}-\rho_{ab})$ together satisfy any fixed (measurable) set of conditions has the limit, as $n \to \infty$, given by the probability that random variables X_{ab} satisfy the same set of conditions, where the X_{ab} are governed by the indicated multivariate normal distribution. To say that R_{ab} *converges in probability* to ρ_{ab} means that the probability that $\rho_{ab} - \epsilon_1 \leq R_{ab} \leq \rho_{ab} + \epsilon_2$ approaches unity as $n \to \infty$, for any positive ϵ_1 and ϵ_2.

[5] Exact distribution theory for the observed maximum frequency in a sample from a multinomial distribution seems intractable. Related material is discussed in [10], [16], and [17].

It would be impracticable to attempt to give here asymptotic distributions relevant to estimators of *each* measure discussed in [8] under *many* sampling methods and *many* assumptions about auxiliary knowledge. Rather, we shall present distributions, for cases that seem important to us, in a way that we hope will enable others to work out similar asymptotic distributions as may be required. Of the various sampling methods, probably the one most frequently found is multinomial sampling over the entire $A \times B$ double polytomy; hence we begin with, and devote most space to, that method.

Although this paper is organized around traditional ideas of hypothesis testing and estimation, the asymptotic distributions presented may also be useful in connection with other approaches to statistical inference, for example, the likelihood-ratio and the neo-Bayesian approaches.

3. MULTINOMIAL SAMPLING OVER THE WHOLE DOUBLE POLYTOMY

3.1. The Index λ_b

An index of association called λ_b, suggested in [8] as appropriate in some situations where asymmetry obtains and order is immaterial, is

$$\lambda_b = \frac{\sum_a \rho_{am} - \rho_{\cdot m}}{1 - \rho_{\cdot m}}. \tag{3.1.1}$$

This measure is the relative decrease in probability of erroneous guessing of B_b (when presented with random individuals) as between A_a unknown and A_a known.

We now discuss the maximum likelihood estimator of λ_b,

$$L_b = \frac{\sum_a R_{am} - R_{\cdot m}}{1 - R_{\cdot m}} = \frac{\sum_a N_{am} - N_{\cdot m}}{n - N_{\cdot m}}. \tag{3.1.2}$$

L_b is defined except when $R_{\cdot m} = 1$. We assume $\rho_{\cdot m} \neq 1$, and by the argument of Section A4 we may neglect, for asymptotic purposes, the possibility that $R_{\cdot m} = 1$. What to do should $R_{\cdot m} = 1$ in a finite sample will be discussed later in this Section.

It is shown in Section A5 that $\sqrt{n}(L_b - \lambda_b)$ is asymptotically normal with mean zero and variance

$$(1 - \sum \rho_{am})(\sum \rho_{am} + \rho_{\cdot m} - 2 \sum{}^r \rho_{am})/(1 - \rho_{\cdot m})^3, \tag{3.1.3}$$

where $\sum^r \rho_{am}$ denotes the sum of the ρ_{am}'s over those values of a such that ρ_{am} is taken on in that column in which $\rho_{\cdot m}$ is taken on. Since we have assumed that no ties exist among the contenders for $\text{Max}_b\, \rho_{ab}$ and $\text{Max}_b\, \rho_{\cdot b}$, the definition of $\sum^r \rho_{am}$ is unambiguous. The variance (3.1.3) is zero if and only if λ_b is zero or one.

To clarify the meaning of $\sum^r \rho_{am}$ we now give a simple example. Suppose the ρ_{ab} table is the 3×4 table

a \ b	1	2	3	4	
1	.14*	.05	.04	.04	.27
2	.04	.18*	.06	.04	.32
3	.04	.05	.24*	.08	.41
	.22	.28	.34	.26	1.00

where the marginal totals appear beyond the double lines. The ρ_{am}'s are indicated by asterisks in their cells; $\rho_{\cdot m}$ is .34, and $\sum^r \rho_{am}$ is .24.

It will also be convenient later to use the notation $\sum^c \rho_{mb}$ to mean the sum of the ρ_{mb}'s over those values of b such that ρ_{mb} is taken on in that row in which $\rho_{m\cdot}$ is taken on. In the above example the ρ_{mb}'s are .14, .18, .24, .08 respectively from left to right, $\rho_{m\cdot}$ is .41, and $\sum^c \rho_{mb} = .24 + .08 = .32$.

We shall also use the notations $\sum^r R_{am}$, $\sum^c R_{mb}$ and $\sum^r N_{am}$, $\sum^c N_{mb}$. They are defined just as above, but for the R_{ab}'s and N_{ab}'s respectively. When we work in terms of the R's or N's, ties may of course exist (although we neglect them for purposes of asymptotic theory) and a procedure for handling them will be suggested in the next Section.

It follows from (3.1.3), in a manner described in Section A5, that the following quantity is asymptotically unit-normal (normal with zero mean and unit variance):

$$\sqrt{n}(L_b - \lambda_b) \sqrt{\frac{(1 - R_{\cdot m})^3}{(1 - \sum R_{am})(\sum R_{am} + R_{\cdot m} - 2 \sum^r R_{am})}}, \quad (3.1.4)$$

under the following assumptions, some of which repeat earlier statements:

i) Multinomial sampling over the entire double polytomy;
ii) ρ_{am}'s and $\rho_{\cdot m}$ unique;
iii) $\rho_{\cdot m} \neq 1$ (i.e., λ_b is well defined); and
iv) $\lambda_b \neq 0$ or 1.

For computational convenience, we give another form of (3.1.4),

$$(L_b - \lambda_b) \sqrt{\frac{(n - N_{\cdot m})^3}{(n - \sum N_{am})(\sum N_{am} + N_{\cdot m} - 2 \sum^r N_{am})}}. \quad (3.1.4a)$$

We note that, to the present level of asymptotic approximation, if $\lambda_b = 0$, then $L_b = 0$. (Indeed, if $\lambda_b = 0$, the probability that $L_b = 0$ has the limit 1.) If $\lambda_b = 1$, then $L_b = 1$ without any asymptotic approximation, providing L_b is well defined.

3.2. *Use of Asymptotic Unit-normality*

For n large, $L_b - \lambda_b$ divided by an ASE is approximately unit-normal. The ASE, which we denote by $g(N_{ab}$'s) to emphasize its dependence on the sample, is the reciprocal of the square root factor in (3.1.4a). Then the probability that $(L_b-\lambda_b)/g(N_{ab}\text{'s})$ lies in an interval (c, d) is approximately $\Phi(d)-\Phi(c)$, where Φ is the unit-normal cumulative distribution function, ubiquitously tabled. For example, the probability that $(L_b-\lambda_b)/g(N_{ab}\text{'s})$ lies between -1.96 and 1.96 is approximately .95. Thus we may readily set up approximate confidence intervals for λ_b. Suppose that we seek an approximate confidence interval, symmetric about L_b, on the $1-\alpha$ level of confidence. Let $K_{\alpha/2}$ be the upper $100(\alpha/2)\%$ point for the unit-normal distribution (e.g., $K_{\alpha/2}=1.96$ for $\alpha=.05$). Then

$$\Pr\{-K_{\alpha/2} \leq (L_b - \lambda_b)/g(N_{ab}\text{'s}) \leq K_{\alpha/2}\} \cong 1 - \alpha, \qquad (3.2.1)$$

or, equivalently,

$$\Pr\{L_b - K_{\alpha/2}g(N_{ab}\text{'s}) \leq \lambda_b \leq L_b + K_{\alpha/2}g(N_{ab}\text{'s})\} \cong 1 - \alpha, \qquad (3.2.2)$$

so that $L_b \pm K_{\alpha/2}g(N_{ab}\text{'s})$ gives a confidence interval approximately on the $1-\alpha$ level of confidence. If the interval happens to go beyond 0 or 1, such inadmissible values would be excluded.

Similarly, we may test the null hypothesis that $\lambda_b=\lambda_b^{(0)}$ (when $\lambda_b^{(0)}\neq 0$),[6] on approximately the α level of significance, by rejecting the null hypothesis just when $\lambda_b^{(0)}$ lies outside of the interval $L_b \pm K_{\alpha/2}g(N_{ab}\text{'s})$.

To test the special hypothesis $\lambda_b=1$, one accepts when $L_b=1$ and otherwise rejects, for any level of significance. (See the final paragraph of Section 3.1.) Because of our level of asymptotic approximation,[6] one accepts the special hypothesis $\lambda_b=0$ when $L_b=0$ and otherwise rejects, for any significance level. Thus, approximate confidence intervals for λ_b, as described above, should exclude the points $\lambda_b=0$ and $\lambda_b=1$ *unless* $L_b=0$ or 1. In the later cases, the confidence interval consists of 0 or 1 (respectively) alone.

One-sided confidence intervals and tests may be readily obtained in the same manner.

It is also possible to obtain confidence intervals for the *difference* between the values of λ_b in two tables. Suppose that we have two independent multinomial samples, one from each of the tables: $\{N_{ab}^{(1)}\}$, $\{N_{ab}^{(2)}\}$. Let $L_b^{(i)}$ and $\lambda_b^{(i)}$ $(i=1, 2)$ be the estimated and true values of λ_b. If the assumptions after (3.14) are satisfied for both tables, then

$$\frac{(L_b^{(1)} - L_b^{(2)}) - (\lambda_b^{(1)} - \lambda_b^{(2)})}{\sqrt{[g(N_{ab}^{(1)}\text{'s})]^2 + [g(N_{ab}^{(2)}\text{'s})]^2}} \qquad (3.2.3)$$

[6] The situation here is much like that of asymptotic distributions for the squared sample correlation coefficient or the squared sample multiple correlation coefficient [2, p. 415]. When the population parameter is zero, the asymptotic distribution appropriate for other cases degenerates and puts all its mass on zero. By changing the power of n as a scaling factor, one can often, in such cases, obtain a nondegenerate limit distribution.

In addition to this parallelism between asymptotic distribution theory for L_b and the squared sample correlation coefficient, there is an interpretive relationship between the two population quantities. If ρ is the population correlation coefficient, then $1-\rho^2$ is the ratio of "unexplained variability" in one variate, when the other is known, to "unexplained variability" when the other is not known. ("Unexplained variability" here refers to expected squared deviation around the best linear predictor and around the best constant predictor, respectively [18, p. 817].) Similarly, $1-\lambda_b$ is the ratio of "unexplained variability" in predicting the B classification, but here measured in terms of error probabilities for prediction from the A classification and from nothing, respectively [8, p. 741].

There are, of course, differences between ρ^2 and λ_b. For example, ρ itself—unsquared—may be positive or negative and its sign gives information about the sense of the association. For λ_b it is meaningless to speak of the sign or sense of association, since λ_b is invariant under permutations of rows (columns) among themselves. Another difference is that ρ is symmetric between the two variates while λ_b is not.

is asymptotically unit-normal. Thus

$$L_b^{(1)} - L_b^{(2)} \pm K_{\alpha/2}\sqrt{[g(N_{ab}^{(1)}\text{'s})]^2 + [g(N_{ab}^{(2)}\text{'s})]^2}$$

gives a confidence interval for $\lambda_b^{(1)} - \lambda_b^{(2)}$ approximately on the $1-\alpha$ level of confidence.

Similarly, we may test the null hypothesis $\lambda_b^{(1)} - \lambda_b^{(2)} = \Delta$, on approximately the α level of significance, by rejecting the null hypothesis just when the above confidence interval for $\lambda_b^{(1)} - \lambda_b^{(2)}$ fails to cover Δ. In particular, we may test the null hypothesis that the difference between $\lambda_b^{(1)}$ and $\lambda_b^{(1)}$ is zero (i.e., that $\lambda_b^{(1)} = \lambda_b^{(2)}$), on approximately the α level of significance, by rejecting this hypothesis when the above confidence interval fails to include zero. This test of the null hypothesis that $\lambda_b^{(1)} = \lambda_b^{(2)}$ can be generalized in order to obtain a test of the null hypothesis that the values of λ_b in k tables are all equal, i.e., that $\lambda_b^{(1)} = \lambda_b^{(2)} = \cdots = \lambda_b^{(k)}$, where $\lambda_b^{(i)}$ is the true value of λ_b for the ith table. First we note that, if the assumptions stated in Section 3.1 are satisfied for each of the k tables, and if the null hypothesis is in fact true, then the statistic

$$\sum_{i=1}^{k} (L_b^{(i)} - \bar{L})^2 / [g(N_{ab}^{(i)}\text{'s})]^2 \tag{3.2.4}$$

will have approximately ($n^{(i)} \to \infty$) the chi-square distribution[7] with $k-1$ degrees of freedom, where $L_b^{(i)}$ ($i = 1, 2, \cdots, k$) denotes the estimated value of λ_b in the ith table, $n^{(i)}$ is the sample size in the ith table, $g(N_{ab}^{(i)}\text{'s})$ denotes the estimate of the asymptotic standard deviation of $L_b^{(i)}$ (the $g(N_{ab}^{(i)}\text{'s})$ are maximum likelihood estimates), and where

$$\bar{L} = \left\{ \sum_{i=1}^{k} \{ L_b^{(i)} [g(N_{ab}^{(i)}\text{'s})]^{-2} \} \right\} \Big/ \left\{ \sum_{i=1}^{k} [g(N_{ab}^{(i)}\text{'s})]^{-2} \right\}.$$

We may test the null hypothesis that $\lambda_b^{(1)} = \lambda_b^{(2)} = \cdots = \lambda_b^{(k)}$, on approximately the α level of significance, by rejecting this hypothesis just when the statistic (3.2.4) is larger than the upper 100α per cent point of the chi-square distribution with $k-1$ degrees of freedom.

We digress to comment briefly on the possibility of using, not L_b, but some monotone transform of it for purposes of hypothesis testing, interval estimation, etc. One considers such transformations in the hope of bettering the asymptotic normal approximation, of simplifying the asymptotic variance, or of making the asymptotic variance more nearly constant. We record here expressions like (3.1.4a) for three transformations; the following three quantities are asymptotically unit-normal:

---

[7] The derivation of this asymptotic chi-square distribution (under the null hypothesis) may be broken into two parts. First, the asymptotic joint distribution of the k quantities like (3.1.4) is shown (via Section A2) to be the same as the corresponding joint distribution, but with true asymptotic variances replacing the sample estimators in (3.1.4). Second, in the asymptotic distribution of these modified quantities, a standard (weighted) average is considered, along with the corresponding standard, weighted sum of squares of residuals. Finally, (3.2.4) and its distribution are obtained by another application of Section A2 in a generalized form.

$$[\log(1 - L_b) - \log(1 - \lambda_b)]\sqrt{\frac{(n - N_{\cdot m})(n - \sum N_{am})}{\sum N_{am} + N_{\cdot m} - 2\sum^r N_{am}}}, \qquad (3.2.5)$$

where the logarithms are to base e,

$$[\sqrt{1 - L_b} - \sqrt{1 - \lambda_b}]2(n - N_{\cdot m})/\sqrt{\textstyle\sum N_{am} + N_{\cdot m} - 2\sum^r N_{am}}, \qquad (3.2.6)$$

$$[\sqrt{L_b} - \sqrt{\lambda_b}][L_b/(1 - L_b)]^{1/2}\, 2(n - N_{\cdot m})/\sqrt{\textstyle\sum N_{am} + N_{\cdot m} - 2\sum^r N_{am}}. \qquad (3.2.7)$$

The stated asymptotic normality follows directly from that of (3.1.4) by an application of Section A3. The methods of using (3.2.5)–(3.2.7) for inference about λ_b are straight-forward analogies to the methods for using (3.1.4).

All the further approximations in this paper may be used in the same ways as those described above. In every case the estimator of the measure of association minus its true value, all divided by a function of the N_{ab}'s, is approximately unit-normal for large n. Hence the above description of statistical procedures will not be repeated each time.

We now give a numerical example of the use of the suggested approximation based on (3.1.4). We drew a random sample of 50 from the population given by the table in Section 3.1.

Random sample of 50 from above population

Numbers are observed N_{ab}'s, and marginal $N_{a\cdot}$'s and $N_{\cdot b}$'s

		b = 1	2	3	4	
a	1	8	5	3	3	19
	2	0	8	1	0	9
	3	0	4	14	4	22
		8	17	18	7	50

$$\sum N_{am} = 8 + 8 + 14 = 30, \quad N_{\cdot m} = 18, \quad \sum^r N_{am} = 14$$

$$L_b = \frac{30 - 18}{50 - 18} = \frac{12}{32} = .3750$$

$$g(N_{ab}\text{'s})^{-1} = \sqrt{\frac{(50 - 18)^3}{(50 - 30)(30 + 18 - 2 \times 14)}} = 9.0510$$

$$1.96/9.0510 = .2166$$

95% approximate confidence interval: $.1584 \leq \lambda_b \leq .5916$.

The confidence interval obtained is rather wide, but on the other hand the sample is not very large relative to the number of cells and their probabilities. The population value of λ_b, 1/3, is covered by the confidence interval for the chosen sample.

When samples are of moderate size, like the above, one often obtains ties among the maximum $N_{\cdot b}$'s and among the maximum N_{ab}'s for some a or a's. Although these ties disappear asymptotically by our assumption of no ties among the true maximum $\rho_{\cdot b}$'s or ρ_{ab}'s, ties must nonetheless be dealt with in real samples. Ties will affect our suggested procedure only via $\sum^r N_{am}$; they will not affect L_b itself, but only $g(N_{ab}$'s). For example, consider the following sample drawn from the same population as above. Here $\sum^r N_{am}$ might be 10 (if column 2's $N_{\cdot 2}=16$ is taken as $N_{\cdot m}$) or 14 (if column 3's $N_{\cdot 3}=16$ is taken as $N_{\cdot m}$). The sample and the two alternative computations are shown below.

Second random sample of 50 from above population

		b: 1	2	3	4	
	1	9	2	1	1	13
a	2	0	10	1	0	11
	3	2	4	14	6	26
		11	16	16	7	50

$$L_b = \frac{33 - 16}{50 - 16} = .5000$$

$$g(N_{ab}\text{'s})^{-1} = \text{either } \sqrt{\frac{34^3}{17\times(33+16-2\times 10)}} \text{ or } \sqrt{\frac{34^3}{17\times(33+16-2\times 14)}}$$

$$= \text{either } 8.9288 \quad \text{or} \quad 10.4926$$

95% approximate confidence interval:

$$\text{either} \quad .2805 \le \lambda_b \le .7195$$

$$\text{or} \quad .3132 \le \lambda_b \le .6868.$$

When ties occur they may be resolved by the flip of a fair coin; this is the method used in the random sampling discussed in Section 3.8. Other methods are possible and perhaps better. For example, one might average the two or more possible values of $g(N_{ab}$'s), or one might take the largest possible value of $g(N_{ab}$'s). This topic requires further investigation.

Another problem that may arise in real samples, even when the assumptions for our asymptotic statements are true, is that $N_{\cdot m}$ may be n (i.e., $R_{\cdot m}=1$). This means that all the observations fall in one column. Unlike ties, this should happen very rarely in the usual sort of application we envisage. There seems to be no reasonable way of estimating λ_b in this case, and in fact L_b is not defined. Thus we suggest that, when $N_{\cdot m}=n$, the confidence interval be taken as the trivial one of all possible values of λ_b: $0 \le \lambda_b \le 1$, and that any null hypothesis be accepted. To give a confidence interval from 0 to 1 inclusive is, of course, just a way of saying that nothing has been learned from the sample about λ_b.

3.3. The Index λ_a

If the roles of columns and rows be interchanged, we have the index λ_a and its estimator L_a. Everything is exactly the same as for λ_b and L_b, except for a systematic interchange of notation.

3.4. The Index λ

In [8], the symmetrical version of λ_a and λ_b was

$$\lambda = \frac{\sum_a \rho_{am} + \sum_b \rho_{mb} - \rho_{\cdot m} - \rho_{m\cdot}}{2 - (\rho_{\cdot m} + \rho_{m\cdot})}. \tag{3.4.1}$$

The maximum likelihood estimator of λ is

$$\begin{aligned} L &= \frac{\sum_a R_{am} + \sum_b R_{mb} - R_{\cdot m} - R_{m\cdot}}{2 - (R_{\cdot m} + R_{m\cdot})} \\ &= \frac{\sum_a N_{am} + \sum_b N_{mb} - N_{\cdot m} - N_{m\cdot}}{2n - N_{\cdot m} - N_{m\cdot}}. \end{aligned} \tag{3.4.2}$$

We assume that λ is determinate, that is, that $\rho_{\cdot m} + \rho_{m\cdot} \neq 2$, or in other words that the entire population does not lie in one cell of the $A \times B$ double polytomy.

We show in Section A6 that (provided λ is determinate and $\neq 0$ or 1) the following quantity is asymptotically unit normal:

$$(L - \lambda)\frac{\sqrt{n}(2 - U_\cdot)^2}{\sqrt{\begin{array}{l}(2 - U_\cdot)(2 - U_\Sigma)(U_\cdot + U_\Sigma + 4 - 2U_*) \\ \quad - 2(2 - U_\cdot)^2(1 - \sum^* R_{am}) - 2(2 - U_\Sigma)^2(1 - R_{**})\end{array}}}, \tag{3.4.3}$$

where

$$\begin{aligned} U_\cdot &= R_{\cdot m} + R_{m\cdot}, \\ U_\Sigma &= \sum R_{am} + \sum R_{mb}, \\ U_* &= \sum{}^r R_{am} + \sum{}^c R_{mb} + R_{*m} + R_{m*}, \end{aligned} \tag{3.4.4}$$

with the asterisked notation defined as follows:

$\sum^* R_{am}$ = sum of those R_{am}'s that also appear as R_{mb}'s, i.e., $\sum\sum R_{ab}$ over all (a, b) such that $R_{ab} = R_{am} = R_{mb}$,
R_{**} = that R_{ab} that appears both in the row for which $R_{a\cdot}$ is maximum and in the column for which $R_{\cdot b}$ is maximum,
R_{*m} = that R_{am} in the same row as that for which $R_{a\cdot}$ is maximum,
R_{m*} = that R_{mb} in the same column as that for which $R_{\cdot b}$ is maximum.

This notation is easier to use than to write down formally. An example of its use follows for the first sample of 50 described in Section 3.2.

First random sample in Section 3.2.

$$\sum N_{am} = 30, \qquad \sum N_{mb} = 34, \qquad N_{\cdot m} = 18, \qquad N_{m\cdot} = 22$$

$$L = \frac{30 + 34 - 18 - 22}{100 - 18 - 22} = .4000$$

$$U_{\cdot} = (18 + 22)/50 = .8000 \qquad U_{\Sigma} = (30 + 34)/50 = 1.2800$$

$$\sum{}^{r} R_{am} = 14/50 = .2800 \qquad \sum{}^{c} R_{mb} = (14 + 4)/50 = .3600$$

$$R_{*m} = 14/50 = .2800 \qquad R_{m*} = 14/50 = .2800$$

$$U_{*} = (14 + 18 + 14 + 14)/50 = 1.200$$

$$\sum{}^{*} R_{am} = (8 + 8 + 14)/50 = .600 \qquad R_{**} = 14/50 = .2800$$

$$g(N_{ab}\text{'s})^{-1} = \frac{\sqrt{50}(2 - .80)^2}{\sqrt{(2 - .80)(2 - 1.28)(.80 + 1.28 + 4 - 2.4) - 2(2 - .80)^2(1 - .60) - 2(2 - 1.28)^2(1 - .28)}}$$

$$\cong 8.9964$$

95% approximate confidence interval for λ: $.1821 \leq \lambda \leq .6179$. Population value of $\lambda = .3600$.

As before, ties may occur that make the various quantities entering into the right fractional factor of (3.4.3) ambiguous, although such ambiguities will disappear asymptotically under our assumptions of unique maxima.

Again, as with L_b, when $\lambda = 0$ or 1, our asymptotic expressions degenerate. If $\lambda = 0$, the square root factor of (3.4.3) becomes 0; to our level of approximation, if $\lambda = 0$, L is 0. If $\lambda = 1$, the population is wholly concentrated in cells no two of which lie in the same row or column; hence the sample will be similarly concentrated, and L will always be 1 without any asymptotic approximation, providing that it is well defined. Thus, as before, if $L \neq 0$ or 1, and if a confidence interval, computed in the described manner, includes 0 or 1 or points beyond 0 or 1, such values should be removed from it. If $L = 0$ or 1, the confidence interval is just the single number 0 or 1 respectively.

If all the observations lie in a single cell, then L is indeterminate. This should be very infrequent for applications of the kind we have in mind. We suggest that, when this occurs, the confidence interval should be the entire interval [0, 1], and that any null hypothesis be accepted.

3.5. The Index γ

In [8], an index of association called γ was suggested as appropriate in some situations where both classifications have intrinsic and relevant order. The definition of γ was

$$\gamma = \frac{\Pi_s - \Pi_d}{1 - \Pi_t} = \frac{2\Pi_s + \Pi_t - 1}{1 - \Pi_t}, \tag{3.5.1}$$

where

$$\Pi_s = 2\sum_a \sum_b \rho_{ab}\Big\{\sum_{a'>b}\ \sum_{b'>b} \rho_{a'b'}\Big\}$$

$$\Pi_d = 2\sum_a \sum_b \rho_{ab}\Big\{\sum_{a'>a}\ \sum_{b'<b} \rho_{a'b'}\Big\}$$

$$\Pi_t = 1 - \Pi_s - \Pi_d = \sum_a \sum_b \rho_{ab}\{\rho_{a\cdot} + \rho_{\cdot b} - \rho_{ab}\}$$

$$= \sum_a \rho_{a\cdot}^2 + \sum_b \rho_{\cdot b}^2 - \sum_a \sum_b \rho_{ab}^2. \tag{3.5.2}$$

Here Π_s ("s" for "same") is the probability that two randomly chosen individuals will have the same order in both classifications (concordance), Π_d ("d" for "different") is the probability that they will have different orders (discordance) and Π_t ("t" for "tie") is the probability that one or both classifications will be the same so that order is not clearly defined. We propose the estimation of γ by its maximum likelihood estimator,

$$G = \frac{P_s - P_d}{1 - P_t} = \frac{2P_s - 1 + P_t}{1 - P_t}, \tag{3.5.3}$$

where the P's are the sample analogs of the Π's, as follows:

$$P_s = 2\sum_a \sum_b R_{ab}\Big\{\sum_{a'>a}\ \sum_{b'>b} R_{a'b'}\Big\}$$

$$= \frac{2}{n^2}\sum_a \sum_b N_{ab}\Big\{\sum_{a'>a}\ \sum_{b'>b} N_{a'b'}\Big\},$$

$$P_d = 2\sum_a \sum_b R_{ab}\Big\{\sum_{a'>a}\ \sum_{b'<b} R_{a'b'}\Big\}$$

$$= \frac{2}{n^2}\sum_a \sum_b N_{ab}\Big\{\sum_{a'>a}\ \sum_{b<'b} N_{a'b'}\Big\},$$

$$P_t = 1 - P_s - P_d = \sum_a \sum_b R_{ab}\{R_{a\cdot} + R_{\cdot b} - R_{ab}\} \tag{3.5.4}$$

$$= \frac{1}{n^2}\sum_a \sum_b N_{ab}\{N_{a\cdot} + N_{\cdot b} - N_{ab}\}$$

$$= \sum_a R_{a\cdot}^2 + \sum_b R_{\cdot b}^2 - \sum_a \sum_b R_{ab}^2$$

$$= \frac{1}{n^2}\Big[\sum_a N_{a\cdot}^2 + \sum_b N_{\cdot b}^2 - \sum_a \sum_b N_{ab}^2\Big].$$

As before, we suppose multinomial sampling over the entire tableau, and we assume $\Pi_t \neq 1$.

It is shown in Section A7 that $\sqrt{n}(G-\gamma)$ is asymptotically normal with zero mean and variance[8]

[8] This variance for the 2×2 case, in which γ is the same as Yule's Q, was obtained by Yule in 1900 [27, p. 285]. For a recent discussion of the 2×2 case, see [22] and [7].

$$\frac{16}{(1-\Pi_t)^4}\left\{\Pi_s^2\Pi_{dd} - 2\Pi_s\Pi_d\Pi_{sd} + \Pi_d^2\Pi_{ss}\right\}, \tag{3.5.5}$$

where

$$\begin{aligned}
\Pi_{ss} &= \sum_a \sum_b \rho_{ab}\Big\{ \sum_{a'>a} \sum_{b'>b} \rho_{a'b'} + \sum_{a'<a} \sum_{b'<b} \rho_{a'b'} \Big\}^2 \\
\Pi_{sd} &= \sum_a \sum_b \rho_{ab}\Big\{ \sum_{a'>a} \sum_{b'>b} \rho_{a'b'} + \sum_{a'<a} \sum_{b'<b} \rho_{a'b'} \Big\} \\
&\qquad \cdot \Big\{ \sum_{a'>a} \sum_{b'<b} \rho_{a'b'} + \sum_{a'<a} \sum_{b'>b} \rho_{a'b'} \Big\}, \\
\Pi_{sd} &= \sum_a \sum_b \rho_{ab}\Big\{ \sum_{a'>a} \sum_{b'<b} \rho_{a'b'} + \sum_{a'<a} \sum_{b'>b} \rho_{a'b'} \Big\}^2.
\end{aligned} \tag{3.5.6}$$

These doubly subscripted Π's are readily interpreted as follows. Suppose we draw three individuals at random and independently from the population in question. Then

- Π_{ss} is the probability that the second and third individuals both have "positive" sign relationships (i.e., are both concordant) with the first.
- Π_{dd} is the probability that the second and third individuals both have "negative" sign relationships (i.e., are both discordant) with the first.
- Π_{sd} is the probability that the second individual has "positive" sign relationships with the first (i.e., the first and second are concordant) but that the third has "negative" sign relationships with the first (i.e., the first and third are discordant).[9]

Hence, if we define P_{ss}, P_{sd}, and P_{dd} just as the corresponding Π's but with R_{ab} replacing ρ_{ab}, and if we assume that (3.5.5) is $\neq 0$, it follows that

$$\sqrt{n}(G-\gamma) \Big/ \sqrt{\frac{16}{(1-P_t)^4}\left\{P_s^2P_{dd} - 2P_sP_dP_{sd} + P_d^2P_{ss}\right\}} \tag{3.5.7}$$

is asymptotically unit-normal. (Note that we do not require any uniqueness of maxima assumptions here. On the other hand, the assumption that (3.5.5)$\neq 0$ does not seem to have any very simple interpretation. Comments on the meaning of this assumption are given in Section A7.)

If $G=1$, then the denominator of (3.5.7) (i.e., the estimator of $\sqrt{(3.5.5)}$) will be equal to zero. For this particular situation, a possible statistical procedure, if n is large, is to give as the confidence interval the degenerate interval $\gamma=1$. If $\gamma<1$, the probability that $G=1$ vanishes as $n\to\infty$, while if $\gamma=1$, G will always be 1, providing that it is well defined. On the other hand, if γ is near 1, G may frequently equal 1 unless n is very large, so that the magnitude of n for our asymptotic theory to work depends critically on γ when γ is near 1. Similar comments can also be made when $G=-1$. In the particular situation where G is undefined, which will happen very rarely in the sort of application we envisage, the point of view presented at the end of Section 3.2 (for the case where L_b is undefined) can be applied.

[9] Note that Π_{sd} has been defined unsymmetrically with respect to the second and third individuals; this is the reason for the 2 in the middle term of the last factor of (3.5.5).

If in computing P_{ss}, etc., we use N_{ab} instead of R_{ab}, we emerge with $n^3 P_{ss}$, etc. Hence, denoting by P_{ss}, P_s, etc., the quantities corresponding to P_{ss}, P_s, etc., but computed in terms of N_{ab}'s, we find that

$$P_{ss} = \mathsf{P}_{ss}/n^3 \qquad P_s = \mathsf{P}_s/n^2, \text{ etc.};$$

that

$$G = \frac{\mathsf{P}_s - \mathsf{P}_d}{n^2 - \mathsf{P}_t} = \frac{\mathsf{P}_s - \mathsf{P}_d}{\mathsf{P}_s + \mathsf{P}_d}\,;$$

and that (3.5.7) is the same as

$$(G - \gamma)\,\frac{(n^2 - \mathsf{P}_t)^2}{4\sqrt{\mathsf{P}_s^2\mathsf{P}_{dd} - 2\mathsf{P}_s\mathsf{P}_d\mathsf{P}_{sd} + \mathsf{P}_d^2\mathsf{P}_{ss}}}\,. \tag{3.5.8}$$

This is perhaps the most convenient form for computation. As an example, let us treat the first random sample used in Section 3.2, but now with each population polytomy thought of as ordered. A convenient way to organize the computations is the following.

First set down the N_{ab} table. Then compute what we might call the S (for "same") table. This $\alpha \times \beta$ table contains in its (a, b) cell the sum of all $N_{a'b'}$ such that $a' > a$ and $b' > b$, plus the sum of all $N_{a'b'}$ such that $a' < a$ and $b' < b$. Then compute the D (for 'different') table. This $\alpha \times \beta$ table contains in its (a, b) cell the sum of all $N_{a'b'}$ such that $a' > a$ and $b' < b$ plus the sum of all $N_{a'b'}$ such that $a' < a$ and $b' > b$. Note that, if we were only computing G itself, we could use simpler S and D tables, entailing but one pair of inequalities.

Thus we have for our example

N_{ab} Table

8	5	3	3
0	8	1	0
0	4	14	4

S Table

31	19	4	0
22	26	17	16
0	8	21	25

D Table

0	0	12	27
11	6	7	18
20	7	3	0

where, e.g., the 31 in the upper left corner of the S table is found by adding $8+1+0+4+14+4=31$, and the 7 in the second row and third column of the D table is found by adding $0+4+3=7$.

P_s is found by multiplying each entry of the N_{ab} table by the corresponding entry of the S table and adding the products.

P_d is found by multiplying each entry of the N_{ab} table by the corresponding entry of the D table and adding the products.

P_{ss} is found by multiplying each entry of the N_{ab} table by the square of the corresponding entry of the S table and adding the products.

P_{dd} is found by multiplying each entry of the N_{ab} table by the square of the corresponding entry of the D table and adding the products.

P_{sd} is found by taking the sum of the triple products of corresponding terms from the N_{ab}, S, and D tables.

With the possible exception of P_{sd}, all these numbers may be found very rapidly with the aid of a table of squares and a desk computer. For our example we have

$$\mathsf{P}_s = 1{,}006, \qquad \mathsf{P}_d = 242$$

$$\mathsf{P}_{ss} = 24{,}168, \qquad \mathsf{P}_{sd} = 2{,}617, \qquad \mathsf{P}_{dd} = 3{,}278.$$

A simple numerical check starts by computation of a T (for 'tie') table. Here in the (a, b) cell one puts $N_{a\cdot}+N_{\cdot b}-N_{ab}$. In our example we have

T Table

19	31	34	23
17	18	26	16
30	35	26	25

(The T table itself may be checked by observing that the sum of its entries must be $n(\alpha+\beta-1)$. In the above case the sum should be 300, as it is.) To use this check compute the following quantities:

P_t = sum of products of N_{ab}'s by corresponding entries in T table,
P_{st} = sum of triple products of N_{ab}'s by corresponding entries in S and T tables,
P_{dt} = sum of triple products of N_{ab}'s by corresponding entries in D and T tables,
P_{tt} = sum of products of N_{ab}'s by *squares* of corresponding entries in T table.

In our case we have

$$\mathsf{P}_t = 1{,}252, \qquad \mathsf{P}_{st} = 23{,}515, \qquad \mathsf{P}_{dt} = 6{,}205, \qquad \mathsf{P}_{tt} = 32{,}880.$$

The following relationships then hold and serve as a partial check:

$$\mathsf{P}_s + \mathsf{P}_d + \mathsf{P}_t = n^2,$$

$$\mathsf{P}_{ss} + 2\mathsf{P}_{sd} + \mathsf{P}_{dd} + 2\mathsf{P}_{st} + 2\mathsf{P}_{dt} + \mathsf{P}_{tt} = n^3.$$

Some other relations that may be used for more detailed checking are

$$n\mathsf{P}_s = \mathsf{P}_{ss} + \mathsf{P}_{sd} + \mathsf{P}_{st},$$

$$n\mathsf{P}_d = \mathsf{P}_{sd} + \mathsf{P}_{dd} + \mathsf{P}_{dt},$$

$$n\mathsf{P}_t = \mathsf{P}_{st} + \mathsf{P}_{dt} + \mathsf{P}_{tt}.$$

These all hold for the computations of our example.

In systematizing the above computations it may be convenient to write separately the cross-product tables of the S and D tables, the S and T tables, and the D and T tables. In our example, these are

$S\times D$

0	0	48	0
242	156	119	288
0	56	63	0

$S\times T$

589	589	136	0
374	468	442	256
0	280	546	625

$D\times T$

0	0	408	621
187	108	182	288
600	245	78	0

From the above numbers we compute that

$$G = \frac{1{,}006 - 242}{1{,}006 + 242} = .6122$$

(to four places), and that, via (3.5.8),

$$(.6122 - \gamma)\frac{1{,}557{,}504}{4\sqrt{3{,}458{,}600{,}992}} = (.6122 - \gamma)6.6209$$

may be considered as an observation from an approximately unit-normal population. Hence we may readily establish approximate confidence limits for γ, say on the 95% level of confidence. These limits, for our sample, say that γ lies between the numbers

$$.6122 \pm \frac{1.96}{6.621},$$

or that γ (to 3 places) lies between .316 and .908. In particular, this means that G differs from zero with statistical significance on the 5% level of significance. (The true value of γ for the cross classification from which the sample was drawn is .4889 to four places.)

The computations described and exemplified in the preceding pages are rather tedious because of the many arithmetical operations. For a rapid significance test or a crude confidence interval, the computations may be much curtailed (at the expense of power).

This curtailment is possible because of the existence of a simple upper bound for the asymptotic variance of $\sqrt{n}\,(G-\gamma)$. We show in Section A7 that the asymptotic variance of $\sqrt{n}\,(G-\gamma)$, whose exact value is given by (3.5.5), is always less than or equal to

$$2(1 - \gamma^2)/(1 - \Pi_t). \tag{3.5.9}$$

This upper bound is closely related to a bound obtained by Daniels and Kendall for a somewhat different problem (see [3], [4]), and to a bound presented by A. Stuart [23] for a measure of association that is similar to our γ [8, pp. 750–1]. The method of obtaining (3.5.9), which is developed in Section A7, is somewhat different from, and perhaps simpler than, the methods used by earlier writers.

As an example of this upper bound, consider the 3×4 population of Section 3.2, from which was drawn the sample we have just discussed. For that population, (3.5.5) turns out to be 1.259 and (3.5.9) to be 2.920, so that the upper bound for asymptotic variance is about 2.3 times the actual asymptotic variance. For most uses, however, the relevant ratio is the square root of 2.3, about 1.5.

It follows from (3.5.9) that *conservative* asymptotic tests and confidence intervals may be obtained by considering as a unit-normal quantity

$$\sqrt{n}(G - \gamma)\sqrt{\frac{1 - P_t}{2(1 - G^2)}}, \tag{3.5.10}$$

or, in another notation,

$$(G - \gamma)\sqrt{\frac{n^2 - \mathsf{P}_t}{2n(1 - G^2)}} \cdot \tag{3.5.11}$$

By a conservative test we mean a test whose probability of falsely rejecting the null hypothesis is known only to be $\leq$ the nominal level of significance. By a conservative confidence interval we mean a confidence interval whose probability of covering the true value being estimated is known only to be $\geq$ the nominal confidence level. This traditional notion of conservatism is, of course, asymmetrical. If we are conservative about significance level, we may lose power.

As an example, let us return to the numerical work a few paragraphs back that led to the 95% confidence interval (.316, .908) from a particular sample. For that sample, using the present cruder approximation,

$$(.6122 - \gamma)\sqrt{\frac{2{,}500 - 1{,}252}{100(1 - .6122^2)}} = (.6122 - \gamma) \times 4.468$$

may be considered as an observation from (approximately) a normal distribution with zero mean and variance less than or equal to unity. This results in the following conservative asymptotic confidence interval at the 95% level of confidence: (.174, 1.051). We would of course change the right-hand end point to obtain (.174, 1.000). (Note that, by a familiar argument, 1.000 itself is excluded from the interval.)

Because of the simple form of (3.5.9), one may consider a variant procedure, much like that of the familiar quadratic confidence procedure for binomial proportions. The probability is (asymptotically) $\geq 1-\alpha$ that

$$-K_{\alpha/2} \leq \sqrt{n}(G - \gamma)\sqrt{\frac{(1 - P_t)}{2(1 - \gamma^2)}} \leq K_{\alpha/2}, \tag{3.5.12}$$

where the quantity in the middle is like (3.5.10) except that G^2 in (3.5.10) is replaced by γ^2 here. The statement (3.5.12) is equivalent to

$$(G - \gamma)^2\left[\frac{n^2 - \mathsf{P}_t}{2n(1 - \gamma^2)}\right] \leq K^2_{\alpha/2}. \tag{3.5.13}$$

Simplifying, we obtain the quadratic inequality

$$\gamma^2[n^2 - \mathsf{P}_t + 2nK^2_{\alpha/2}] - \gamma[2G(n^2 - \mathsf{P}_t)] + G^2(n^2 - \mathsf{P}_t) - 2nK^2_{\alpha/2} \leq 0. \tag{3.5.14}$$

It is readily shown that, for large n, the probability is nearly one that the values of γ satisfying (3.5.14) form an interval. It is a conservative approximate confidence interval for γ.

In our numerical example, (3.5.14) becomes

$$1{,}632\gamma^2 - 1{,}528\gamma + 83.6 \leq 0,$$

and the two real roots of the above quadratic expression are (.058, .878).

The numerical results from our particular sample may be recapitulated as follows:

95% asymptotic confidence interval based on estimate of variance via ((3.5.8): (.316, .908)

95% conservative asymptotic confidence interval based on estimated variance bound via (3.5.11): (.174, 1.000)

95% conservative asymptotic confidence interval based on estimated variance bound via (3.5.14): (.058, .878)

As would be expected, the confidence interval obtained via (3.5.8) is appreciably narrower than the other two. This reflects the fact that the upper bound for asymptotic variance (3.5.9) may be considerably larger than the true asymptotic variance. The two conservative methods give intervals of about the same length but in different positions. If we compare the uncurtailed interval based on (3.5.11), that is (.174, 1.051), with the interval based on (3.5.14), we see that the former is longer than the latter. This might have been expected since G in the denominator of (3.5.11) is subject to sampling variability, while in (3.5.14) the true value γ appears instead of G.

On the whole, we recommend the use of (3.5.8) because of its more precise results. It is true that it requires a tedious (although not difficult) computation, but in most serious studies this amount of computation would be a negligible cost unless it had to be repeated many times. It may well be that better bounds than (3.5.9) will be found that permit simplified computations without appreciable loss of precision. Daniels [4] has shown that the upper bound first given by Daniels and Kendall [3], which is related to (3.5.9) though appropriate for a somewhat different problem, is in general a poor one, although it is sometimes attainable. (See also [14].) Nevertheless, when n is large, even this bound can be good enough for some practical purposes [23], which suggests that the bound (3.5.9) presented here can also be good enough for such practical purposes.

We mentioned earlier in this section that the upper bound (3.5.9) for the variance of G was closely related to a bound presented by Stuart [23] for an estimate of a measure of association that he has suggested. It was also noted in [8, pp. 750–1] that Stuart's measure of association was closely related to γ. (The denominator $1-\Pi_t$ in (3.5.1) does not appear in Stuart's measure, and in its place we find a quantity that depends on the minimum number m of rows and columns in the cross classification table [8].)[10] It therefore seemed worthwhile to include here some numerical comparison of the two measures and the

[10] Stuart's denominator is introduced in order that his measure of association, τ_c, may attain, or nearly attain, the absolute value 1 when the entire cross classification population lies in a longest diagonal. The absolute value 1 is attained, following Stuart, just when the following three conditions are met: (1) population size is a multiple of m, (2) the population lies entirely in cells along a longest diagonal of the cross classification table, and (3) the frequencies in these diagonal cells are equal. This characteristic of τ_c is rather different from the corresponding characteristic of γ, which has absolute value 1 when (but not only when) the population is concentrated on *any* diagonal of the cross classification; in particular, the cells of the diagonal need not have equal or nearly equal frequencies. This difference between τ_c and γ may be relevant in deciding which measures to use in applications.

Stuart defines τ_c for a finite population and considers sampling without replacement, while we consider sampling with replacement. Sampling theory for either τ_c or γ could, of course, be considered for sampling both with and without replacement.

bounds for their corresponding variances. For this purpose, the data presented by Stuart [23, pp. 8–10] has been re-examined. Stuart's data (Table 1 in [23]) refers to 4×4 cross classifications between left and right eye vision for a group of employees in Royal Ordnance factories; he gives two cross classifications, one for men and one for women. The following table summarizes information about t_c, G, and estimated bounds on standard error for Stuart's data:

	t_c	G	Estimated Bound on Standard Error	
			for t_c	for G
Men	+0.629	+0.776	0.029	0.022
Women	+0.633	+0.798	0.019	0.014
			For differences	
Difference	0.004	0.022	0.035	0.027

We note that the estimates G are larger than the corresponding estimates t_c and the estimated bounds on standard errors are smaller. These differences no doubt reflect in part the fact that t_c and G estimate somewhat different population measures of association.

In closing this section, we mention that, in testing the null hypothesis $\gamma=0$, some modification and simplification of the asymptotic variance formula (3.5.8) is possible, since $\Pi_s=\Pi_d$ when the null hypothesis is, in fact, true. In this situation, the denominator of (3.5.8) might be replaced by the following, which is asymptotically equivalent to that denominator under the null hypothesis:

$$2(\mathsf{P}_s + \mathsf{P}_d)\sqrt{\mathsf{P}_{dd} - 2\mathsf{P}_{sd} + \mathsf{P}_{ss}}. \tag{3.5.15}$$

The statistic (3.5.15) is simpler to compute than the denominator of (3.5.8), and its use as a replacement for the denominator of (3.5.8) will not greatly affect, when n is large, the level of significance of the correspondingly modified test, although it will affect, to an unknown extent, the power of this test. Further study of the use of (3.5.15), when the null hypothesis is $\gamma=0$, would be worthwhile. (A related discussion in the binomial context is given in [20].)

If we wished to test the stricter null hypothesis of independence between the two polytomies (independence implies that $\gamma=0$, but $\gamma=0$ does not imply independence), further modification and simplification is possible, in the sense that (3.5.15) could be replaced by a statistic which would be a function of only the marginal N's and which would be asymptotically equivalent to (3.5.15) under the null hypothesis.

3.6. Measures of Reliability

In [8], a number of measures of reliability or agreement were suggested as possibly useful when A_1 is the same class as B_1, A_2 as B_2, and so on, but where assignment to class is by two different methods. In these situations $\alpha=\beta$.

For some of the proposed measures of reliability, namely

$$\sum_{a=1}^{\alpha} \rho_{aa} \quad \text{and} \quad \sum_{|a-b|\leq I} \rho_{ab},$$

where I is a small integer, there is relatively little difficulty in working with the sampling distributions of the sample analogs

$$\sum_{a=1}^{\alpha} R_{aa} \quad \text{and} \quad \sum_{|a-b|\leq I} R_{ab}. \tag{3.6.1}$$

Each of these will, under over-all multinomial sampling, have a binomial distribution with binomial probability equal to the population value of the measure and sample size equal to n. Thus familiar procedures for estimating and testing binomial probabilities may be used.

A slightly more complex measure, of possible interest in the unordered case, was also suggested in [8]. It is

$$\lambda_r = \frac{\sum \rho_{aa} - \frac{1}{2}(\rho_{M\cdot} + \rho_{\cdot M})}{1 - \frac{1}{2}(\rho_{M\cdot} + \rho_{\cdot M})}, \tag{3.6.2}$$

where

$$\rho_{M\cdot} + \rho_{\cdot M} = \operatorname*{Max}_a (\rho_{a\cdot} + \rho_{\cdot a}).$$

The sample analogue of λ_r is

$$L_r = \frac{\sum R_{aa} - \frac{1}{2}(R_{M\cdot} + R_{\cdot M})}{1 - \frac{1}{2}(R_{M\cdot} + R_{\cdot M})}, \tag{3.6.3}$$

where

$$R_{M\cdot} + R_{\cdot M} = \operatorname*{Max}_a (R_{a\cdot} + R_{\cdot a}).$$

Assume now (i) that λ_r is well-defined, (ii) that there is a unique modal class (i.e., that $\rho_{a\cdot}+\rho_{\cdot a}=\rho_{M\cdot}+\rho_{\cdot M}$ for only one a), and (iii) that $\lambda_r \neq \pm 1$. Then the methods of the preceding sections and the Appendix may be applied to show that

$$(L_r - \lambda_r)(1 - \tfrac{1}{2}S_r)^2 \left[\frac{1 - D_r}{n} \{ D_r + \tfrac{1}{4}S_r(1 - D_r - S_r) - R_{MM}(\tfrac{3}{2} + \tfrac{1}{2}D_r - S_r)\} \right]^{-1/2}$$

$$= (L_r - \lambda_r)(1 - \tfrac{1}{2}S_r)^2 \left[\frac{1 - D_r}{n} \{(1 - \tfrac{1}{2}S_r)(\tfrac{1}{2}S_r + D_r - 2R_{MM}) - \tfrac{1}{4}(1 - D_r)(S_r - 2R_{MM})\} \right]^{-1/2} \tag{3.6.4}$$

is asymptotically unit-normal. The new notation is defined as follows:

$$D_r = \sum R_{aa}, \qquad S_r = R_{M\cdot} + R_{\cdot M},$$
$$R_{MM} = \text{that } R_{aa} \text{ such that } R_{a\cdot} + R_{\cdot a} = R_{M\cdot} + R_{\cdot M}. \tag{3.6.5}$$

For large n, R_{MM} is uniquely defined with high probability. If a tie occurs we suggest a random choice.

3.7. Partial and Multiple Association

As final examples of asymptotic sampling theory in the case of fully multinomial sampling, we consider one of the coefficients of partial association suggested in Section 11 of [8] for a three-way classification, together with the coefficient of multiple association suggested in Section 12 of [8].

Suppose that there are three polytomies: $A_1, \cdots, A_\alpha$; $B_1, \cdots, B_\beta$; and $C_1, \cdots, C_\gamma$. If an individual is chosen at random from the fixed triply polytomous population of interest, the probability that he will simultaneously fall in the categories A_a, B_b, and C_c is ρ_{abc}. A measure of partial association between the A and B polytomies, averaged over the C polytomy, is

$$\lambda_b'(A, B \mid C) = \frac{\sum_a \sum_c \rho_{amc} - \sum_c \rho_{\cdot mc}}{1 - \sum_c \rho_{\cdot mc}}, \tag{3.7.1}$$

where dots mean summation over the dotted subscript, where $\rho_{amc} = \text{Max}\ \rho_{abc}$, and where $\rho_{\cdot mc} = \text{Max}\ \rho_{\cdot bc}$. This measure is the relative decrease in probability of error of guessing the B category of an individual if we know both his A and C categories as against knowing only his C category. Thus it refers to optimal prediction with the C category always known (hence "partial" association). It is asymmetric in that only prediction of B categories are considered, and it is unchanged by independent permutations of the classes within each polytomy (hence appropriate in some situations where there is no natural ordering of these classes).

For convenience we may simply write λ_b' instead of λ_b' $(A, B \mid C)$ in this Section, but in applications the arguments should probably be retained, for the six possible asymetrical λ''s obtained from an $\alpha \times \beta \times \gamma$ cross classification will in general all have both different numerical values and different interpretations.

As before, we assume that all relevant maxima are unique, i.e., that

For each a and c, $\rho_{amc} = \rho_{abc}$ for a unique value of b.

For each c, $\rho_{\cdot mc} = \rho_{\cdot bc}$ for a unique value of b.

We also assume that λ_b' is well defined, i.e., that $\sum_c \rho_{\cdot mc} \neq 1$.

Suppose that a sample of n is drawn from our population of interest and that each member of the sample is assigned to one of the $\alpha\beta\gamma$ cells (A_a, B_b, C_c) according to observation (without error) of its categories in the three polytomies. Suppose further that sampling is with replacement, or, alternatively, that the population of interest is infinite or very large. Let N_{abc} be the number of individuals among the n in the sample that fall into the (A_a, B_b, C_c) cell. Then $\sum_a \sum_b \sum_c N_{abc} = n$ and the N_{abc}'s have jointly a multinomial distribution with cell probabilities ρ_{abc}. Denote by R_{abc} the quantity N_{abc}/n.

Just as in Section 2, the $\alpha\beta\gamma$ random variables $\sqrt{n}\ (R_{abc} - \rho_{abc})$ are jointly asymptotically normal, and we may assume, for asymptotic purposes, that R_{amc} and $R_{\cdot mc}$ are taken on at the values of b corresponding to ρ_{amc} and $\rho_{\cdot mc}$.

In this context, we now discuss the approximate distribution of the maximum likelihood estimator of λ_b',

$$L_b' = \frac{\sum_a \sum_c R_{amc} - \sum_c R_{\cdot mc}}{1 - \sum_c R_{\cdot mc}}. \tag{3.7.2}$$

Of course, even if λ_b' is defined, L_b' may not be, but, as in Sections 3.1 and 3.2 we may neglect this possibility for asymptotic purposes. As in prior sections, one may show that $\sqrt{n}\,(L_b' - \lambda_b')$ is asymptotically normal with zero mean and variance

$$(1 - \sum_c \sum_a \rho_{amc})(\sum_c \sum_a \rho_{amc} + \sum_c \rho_{\cdot mc} - 2 \sum_c \sum_a{}^r \rho_{amc})/(1 - \sum_c \rho_{\cdot mc})^3, \tag{3.7.3}$$

where

$$\sum_a{}^r \rho_{amc}$$

denotes the sum of the ρ_{amc}'s (for fixed c) over these values of a such that ρ_{amc} is taken on for that value of b for which $\rho_{\cdot mc}$ is taken on. Note the strong similarity between (3.1.3) and (3.7.3); the latter is just like the former except that all terms (but unity) have an additional summation over c. We note that (3.7.3) is zero if and only if λ_b' is zero or one.

Hence, the following quantity is asymptotically unit-normal:

$$\sqrt{n}[L_b'(A, B \mid C) - \lambda_b'(A, B \mid C)] \cdot \sqrt{\frac{(1 - \sum_c R_{\cdot mc})^3}{(1 - \sum_c \sum_a R_{amc})(\sum_c \sum_a R_{amc} + \sum_c R_{\cdot mc} - 2 \sum_c \sum_a{}^r R_{amc})}} \tag{3.7.4}$$

provided that our earlier assumptions hold and that $\lambda_b' \neq 0$ or 1. If $\lambda_b' = 0$, then, to the present level of asymptotic approximation, $L_b' = 0$. If $\lambda_b' = 1$, then $L_b' = 1$ without any asymptotic approximation, providing L_b' is defined.

We note that the following quantities, are also asymptotically unit-normal:

$$\sqrt{n}[\log(1 - L_b') - \log(1 - \lambda_b')]\sqrt{(1 - \sum_c R_{\cdot mc})(1 - \sum_c \sum_a R_{amc})/D}, \tag{3.7.5}$$

$$\sqrt{n}[\sqrt{1 - L_b'} - \sqrt{1 - \lambda_b'}]2(1 - \sum_c R_{\cdot mc})/\sqrt{D}, \tag{3.7.6}$$

$$\sqrt{n}[\sqrt{L_b'} - \sqrt{\lambda_b'}][L_b'/(1 - L_b')]^{1/2} 2(1 - \sum_c R_{\cdot mc})/\sqrt{D}, \tag{3.7.7}$$

where

$$D = \sum_c \sum_a R_{amc} + \sum_c R_{\cdot mc} - 2 \sum_c \sum_a{}^r R_{amc},$$

and logarithms are to base e.

We conclude this Section with some comments about the following measure of multiple association between B and (A, C) together [8, Sec. 12]:

$$\lambda_b''(B; A, C) = \frac{\sum_a \sum_c \rho_{amc} - \rho_{\cdot m \cdot}}{1 - \rho_{\cdot m \cdot}},$$

where $\rho_{\cdot m \cdot} = \text{Max}\, \rho_{\cdot b \cdot}$. This quantity is exactly λ_b itself, computed from the $(\alpha\gamma) \times \beta$ cross classification in which the $\alpha\gamma$ cells of the A, C cross classification are thought of as making a single classification. (Note: On p. 762 of [8], the rearranged tableau is shown for A against (B, C) rather than for B against (A, C).) The interpretation of $\lambda_b''(B; A, C)$ is the usual one for λ_b, but now comparing errors in predicting B, knowing A *and* C against knowing nothing.

It is interesting to note on analogy (brought to our attention by J. Mincer) with classical correlation analysis. In the present context, the marginal λ_b, relative to prediction of B from C, is

$$\lambda_b(B; C) = \frac{\sum_c \rho_{\cdot mc} - \rho_{\cdot m \cdot}}{1 - \rho_{\cdot m \cdot}},$$

so we see that

$$1 - \lambda_b''(B; A, C) = [1 - \lambda_b(B; C)][1 - \lambda_b'(B, A \mid C)].$$

This is completely analogous to the classical relationship between multiple and partial correlation coefficients,

$$1 - R_{b\cdot ac}^2 = [1 - \rho_{bc}^2][1 - \rho_{ba\cdot c}^2];$$

see, for example, [2], p. 307. It is interesting to note that, although $R_{b\cdot ac}$ may be expressed as a function of ρ_{ba}, ρ_{bc}, and ρ_{ac}, the analogous relationship does not hold for $\lambda_b'(B, A \mid C)$.[11]

Finally, we note that the asymptotic distribution of the sample analog of λ_b'' needs no fresh discussion here. For λ_b'' is really λ_b for the cross classification between (say) B and (A, C); hence the material of Section 3.1 applies directly.

3.8. Sampling Experiments

This section presents the results of some sampling experiments that bear on the adequacy of the asymptotic approximations given in the prior sections.

Table 3.8.1 describes the sampling experiments that have been done; in each case, reference is made to the figure that summarizes the results of the experiment graphically. These figures are drawn on normal probability paper, the straight lines represent the standard normal distribution, and each dot has its abscissa equal to the value of a computed statistic and its ordinate equal to the proportion of computed statistics less than or equal to the abscissa. Thus the dots give the "corners" of the observed sample cumulative distribution functions. Deviations of the dots from the straight line arise from two sources: (1) inadequacy of the asymptotic approximation, and (2) sampling fluctuations.

[11] The simple relationships between the ρ's are inherent in the geometry of classical correlation theory, and are not tied to the assumption of normality. The last sentence of Section 12 in [8] is corrected by this remark.

TABLE 3.8.1. SAMPLING EXPERIMENTS

Sample Measure of Association	Statistic Computed from Each Sample	Sample Size	Number of Samples	Population	Figure
L_b	(3.1.4)	200	50	3×4 cross classification of Sec. 3.2	3.8.1
L_b	(3.1.4)	100	50	3×4 cross classification of Sec. 3.2	3.8.2
L_b	(3.1.4)	100	100 (iv)	3×4 cross classification of Sec. 3.2	3.8.3
L_a (i)	(3.3.4)	200	50	3×4 cross classification of Sec. 3.2	3.8.4
L_a	(3.3.4)	100	50	3×4 cross classification of Sec. 3.2	3.8.5
L_a	(3.3.4)	100	100 (iv)	3×4 cross classification of Sec. 3.2	3.8.6
L_b	(3.1.4)	200	50	2×3 cross classification (ii)	3.8.7
L_a (i)	(3.3.4)	200	50	2×3 cross classification (ii)	3.8.8
G	$\frac{\sqrt{n}\,(G-\gamma)}{\sqrt{(3.5.5)}}$ (iii)	50	100	3×4 cross classification of Sec. 3.2	3.8.9
G	$\frac{\sqrt{n}\,(G-\gamma)}{\sqrt{(3.5.5)}}$	200	50	3×4 cross classification of Sec. 3.2	3.8.10
G	$\frac{\sqrt{n}\,(G-\gamma)}{\sqrt{(3.5.5)}}$	200	50	2×3 cross classification (ii)	3.8.11

Notes to the table

(i) L_a for the 3×4 cross classification is, of course, L_b for the transposed (4×3) cross classification. In fact, the same samples were used.

(ii) The 2×3 cross classification used here was

		b: 1	2	3	
a	1	.04	.23	.18	.45
	2	.18	.05	.32	.55
		.22	.28	.50	1.00

For this population, $\lambda_b = .10$, $\lambda_a = .40$, and $\lambda \approx .03$.

(iii) This statistic is the deviation of G from its population value, γ, divided by the *population* asymptotic std. dev. of G. Thus the results here are not directly applicable, in general, since (3.5.5) would hardly ever be known in practice. The corresponding practical statistic, (3.5.7), requires considerably more computation; sampling results for it are presented in Table 3.8.2.

(iv) The 100 samples include the 50 samples of the line just above.

TABLE 3.8.2. THE EMPIRICAL AVERAGE NUMBER OF TIMES PER ONE HUNDRED SAMPLES THAT $\sqrt{n}\ (G-\gamma)$ DIVIDED BY THE SQUARE ROOT OF (3.5.5) EXCEEDED THE VALUES FROM THE UNIT-NORMAL DISTRIBUTION AT THE TWO-SIDED .05, .10, AND .25 LEVELS, WHEN THE SAMPLE SIZE $n=50$

Values of γ	−.01 to +.01	.20 to .29	.30 to .39	.40 to .49	.50 to .59	.60 to .69	.70 to .79	.80 to .89	.90 to .92	.99 to 1.00
Number of samples	300	800	1900	700	1000	2000	1400	700	300	900
Significance levels										
.25	22	29	26	25	30	30	31	30	26	11
.10	10	13	11	9	12	12	12	9	8	6
.05	6	7	5	4	5	5	5	4	3	5

THE EMPIRICAL AVERAGE NUMBER OF TIMES PER ONE HUNDRED SAMPLES THE STATISTIC (3.5.7) EXCEEDED THE VALUES FROM THE UNIT-NORMAL DISTRIBUTION AT THE TWO-SIDED .05, .10, AND .25 LEVELS, WHEN THE SAMPLE SIZE $n=50$

Values of γ	−.01 to +.01	.20 to .29	.30 to .39	.40 to .49	.50 to .59	.60 to .69	.70 to .79	.80 to .89	.90 to .92	.99 to 1.00
Number of samples	300	800	1900	700	1000	2000	1400	700	300	900
Significance levels										
.25	23	30	29	25	34	33	37	36	32	67
.10	12	17	15	13	18	21	23	24	21	65
.05	6	10	10	7	12	16	18	19	17	65

Note that for the cross classifications considered, with 6 or 12 cells, sample sizes of 50 and 100 would not ordinarily be considered large. The sample is spread over all cells, while, at least for L_b and L_a, only some cells determine the sample measure of association.

In the light of these considerations, we find the sampling results encouraging. For L_a and L_b, the dots on the figures lie generally near the standard normal straight line. The tail probabilities are particularly important for most statistical applications, and we note that some of the deviations in the tails appear large. In examining the figures, however, keep in mind that the scale near the tails is magnified because of the use of normal probability paper.

We next summarize in Table 3.8.2 some sampling results about G that were obtained by Miss Irene Rosenthal (Institute of Child Welfare, University of California, Berkeley) and that we reproduce here with her kind permission.[12]

Miss Rosenthal's work related to two statistics $\sqrt{n}\ (G-\gamma)/\sqrt{(3.5.5)}$ and (3.5.7). She considered a large number of 5×5 cross classifications, categorized by their values of γ. In Table 3.8.2, the populations are not separated, but results are grouped by ranges of γ values. Table 3.8.2 compares the two-sided tail probabilities from asymptotic unit normality with the observed relative fre-

[12] Miss Rosenthal has in preparation a manuscript giving more detailed information about her sampling experiments.

FIG. 3.8.1. Results of Sampling Experiment for (3.1.4). 50 Samples, Each of 200 Cross Classified Observations.

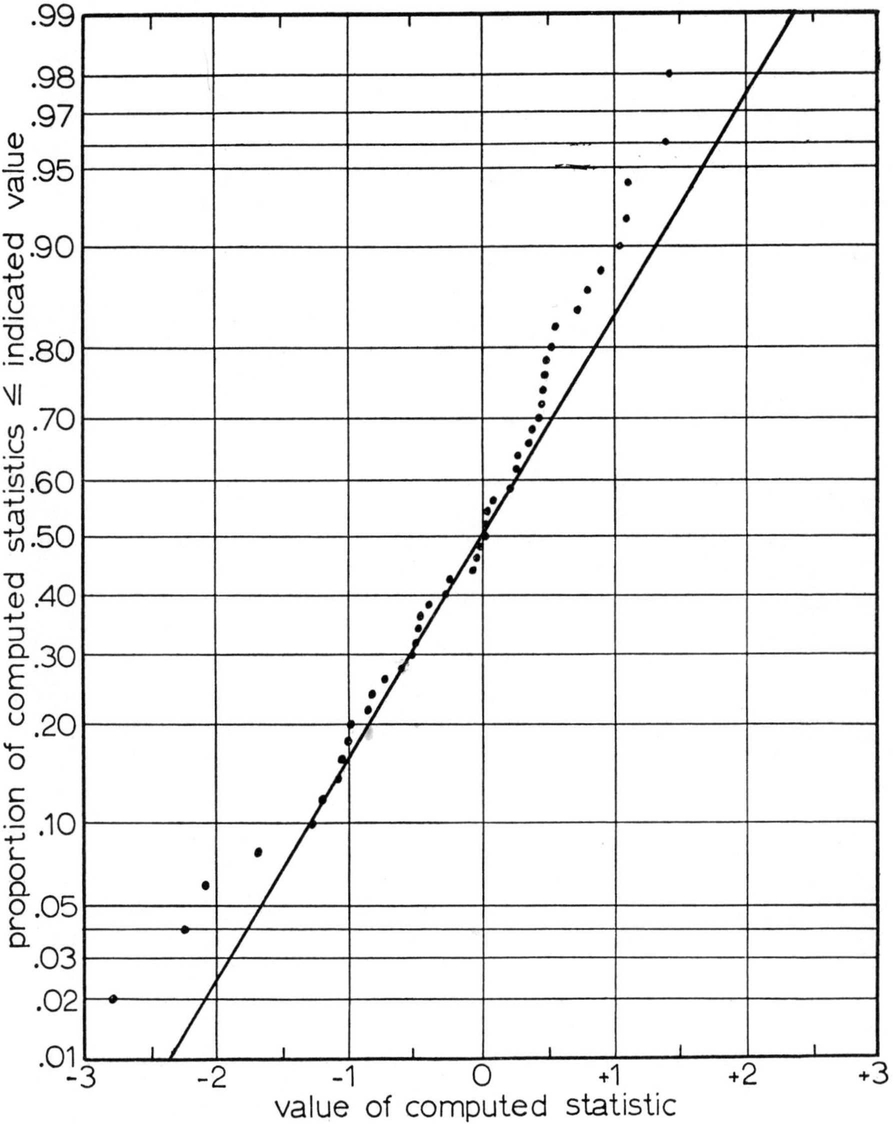

Straight line is unit-normal cumulative.
Unplotted maximum observed value is 1.64.

Fig. 3.8.2. Results of Sampling Experiment for (3.1.4). 50 Samples, Each of 100 Cross Classified Observations.

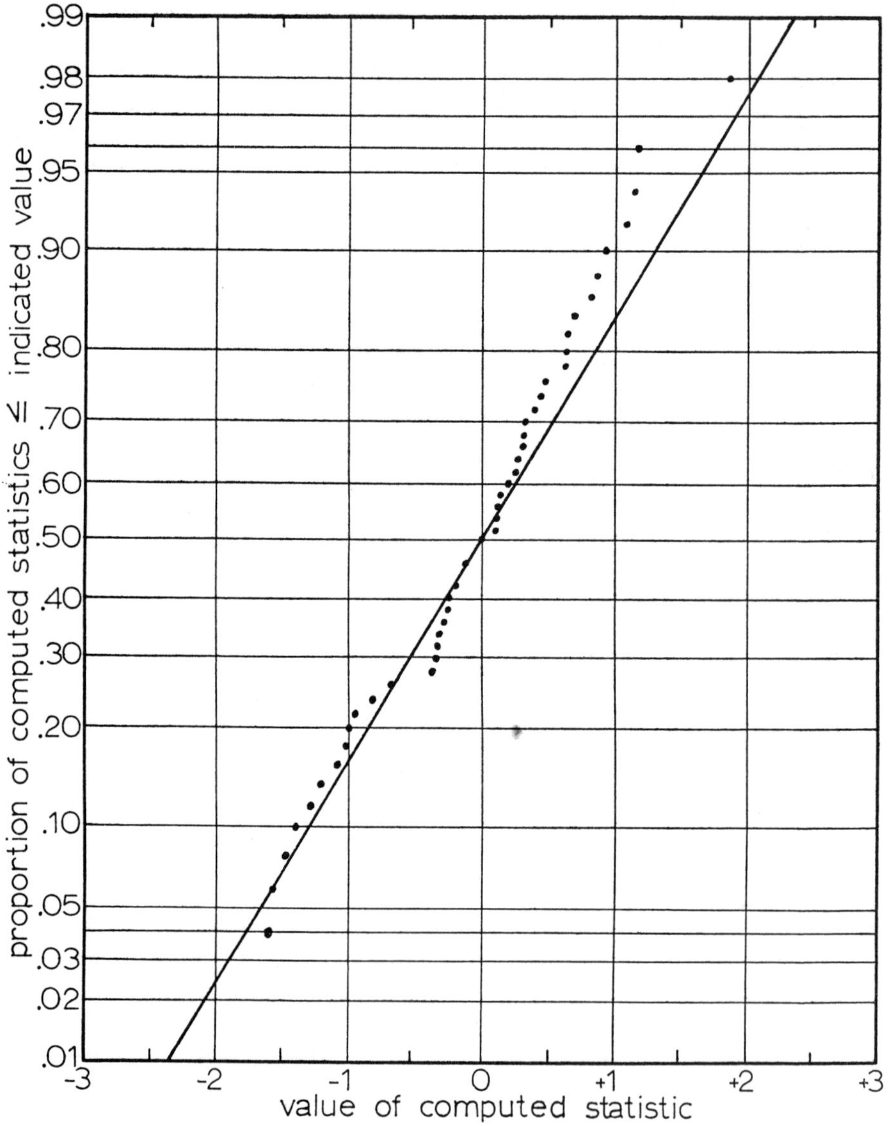

Straight line is unit-normal cumulative.
Unplotted maximum observed value is 2.05; minimum −4.75.

FIG. 3.8.3. Results of Sampling Experiment for (3.1.4). 100 Samples, Each of 100 Cross Classified Observations.

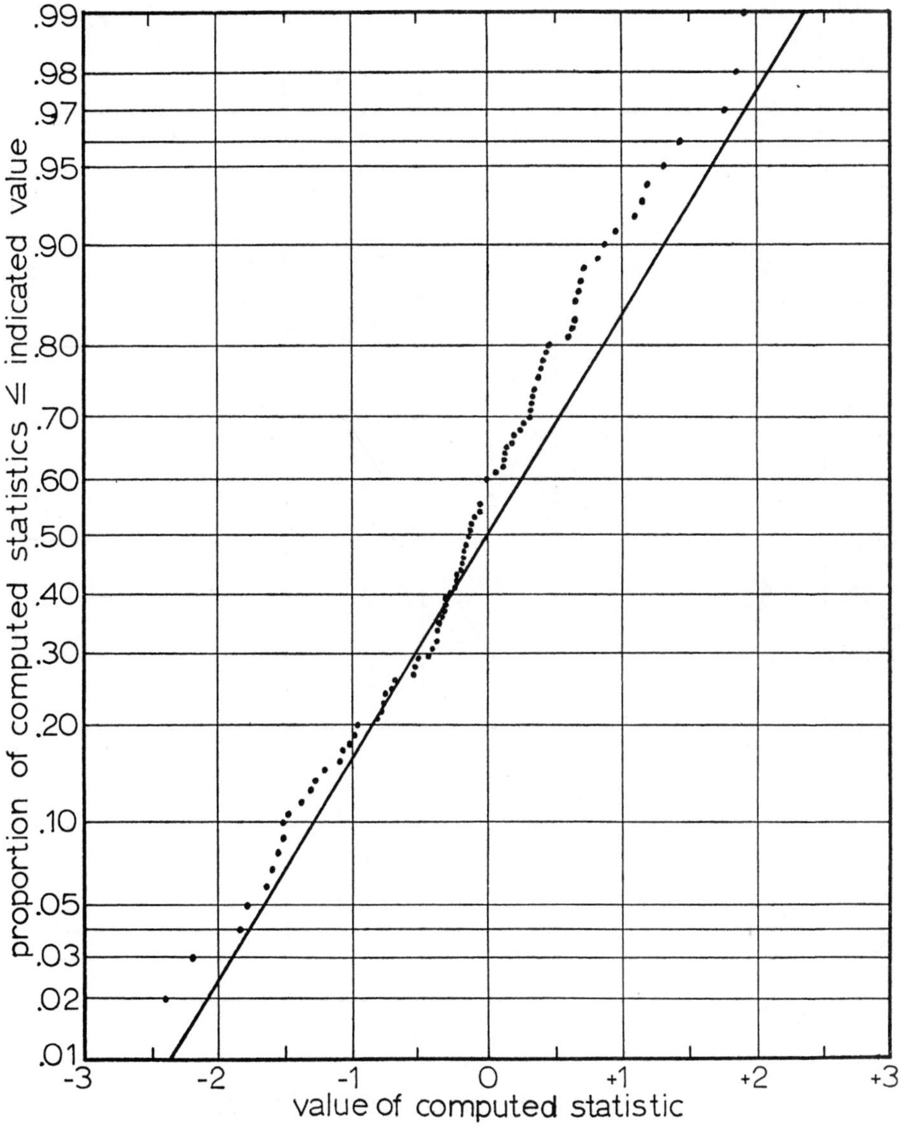

Straight line is unit-normal cumulative.
Unplotted maximum observed value is 2.05; minimum −4.75.

FIG. 3.8.4. Results of Sampling Experiment for (3.3.4). 50 Samples, Each of 200 Cross Classified Observations.

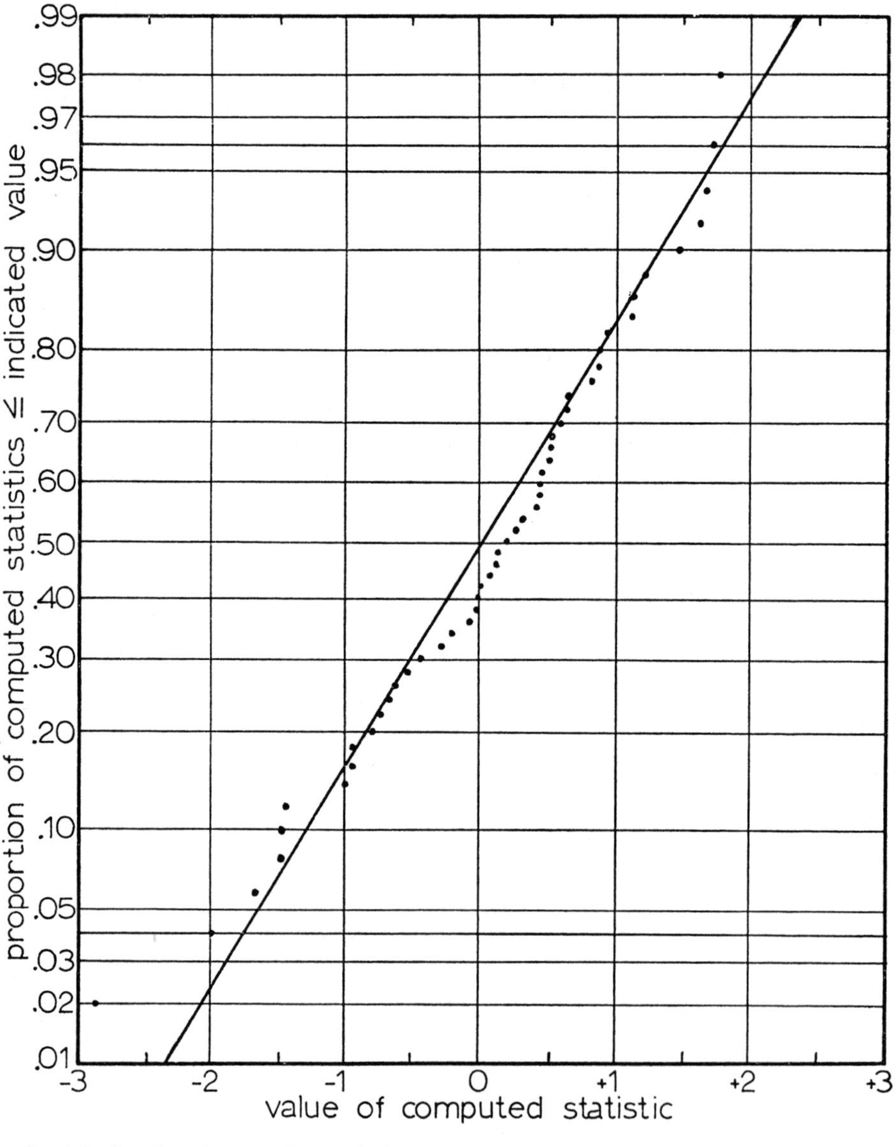

Straight line is unit-normal cumulative.
Unplotted maximum observed value is 2.33.

FIG. 3.8.5. Results of Sampling Experiment for (3.3.4). 50 Samples, Each of 100 Cross Classified Observations.

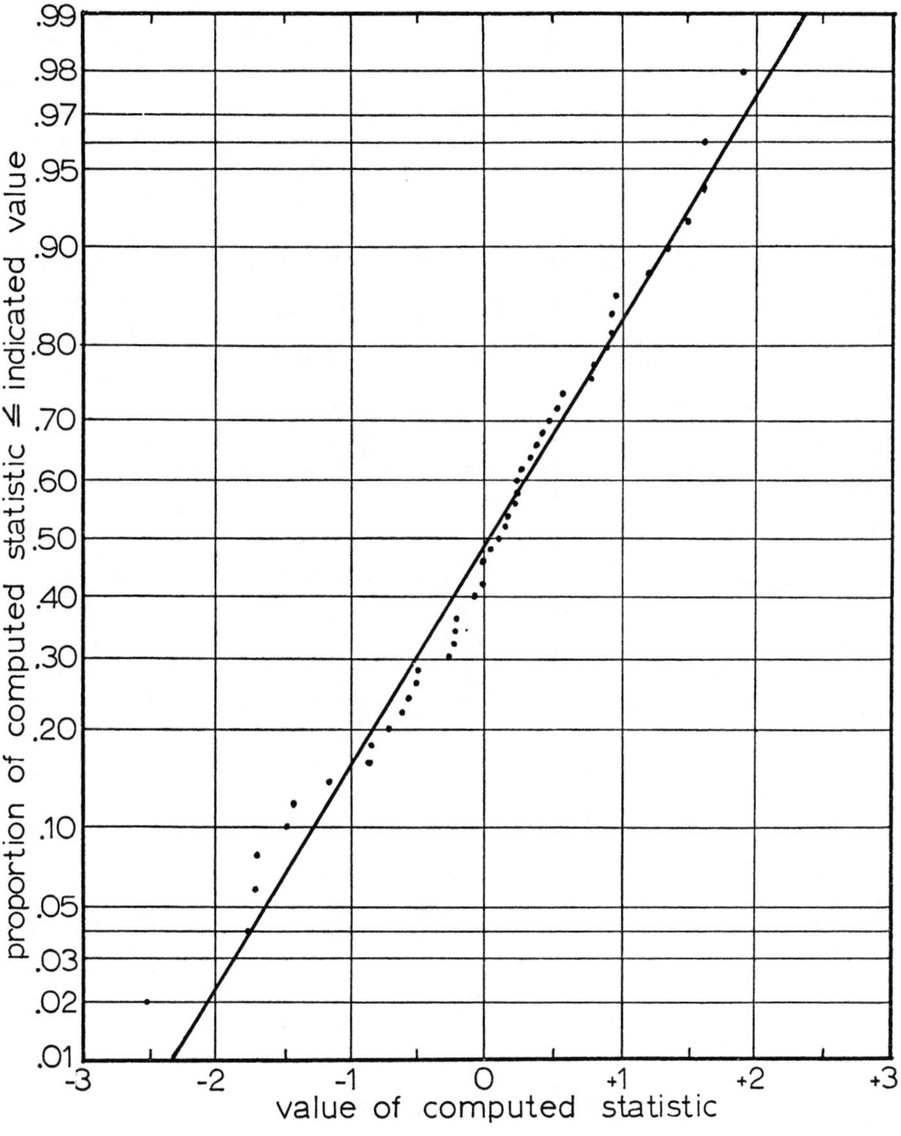

Straight line is unit-normal cumulative.
Unplotted maximum observed value is 2.19.

FIG. 3.8.6. Results of Sampling Experiment for (3.3.4). 100 Samples, Each of 100 Cross Classified Observations.

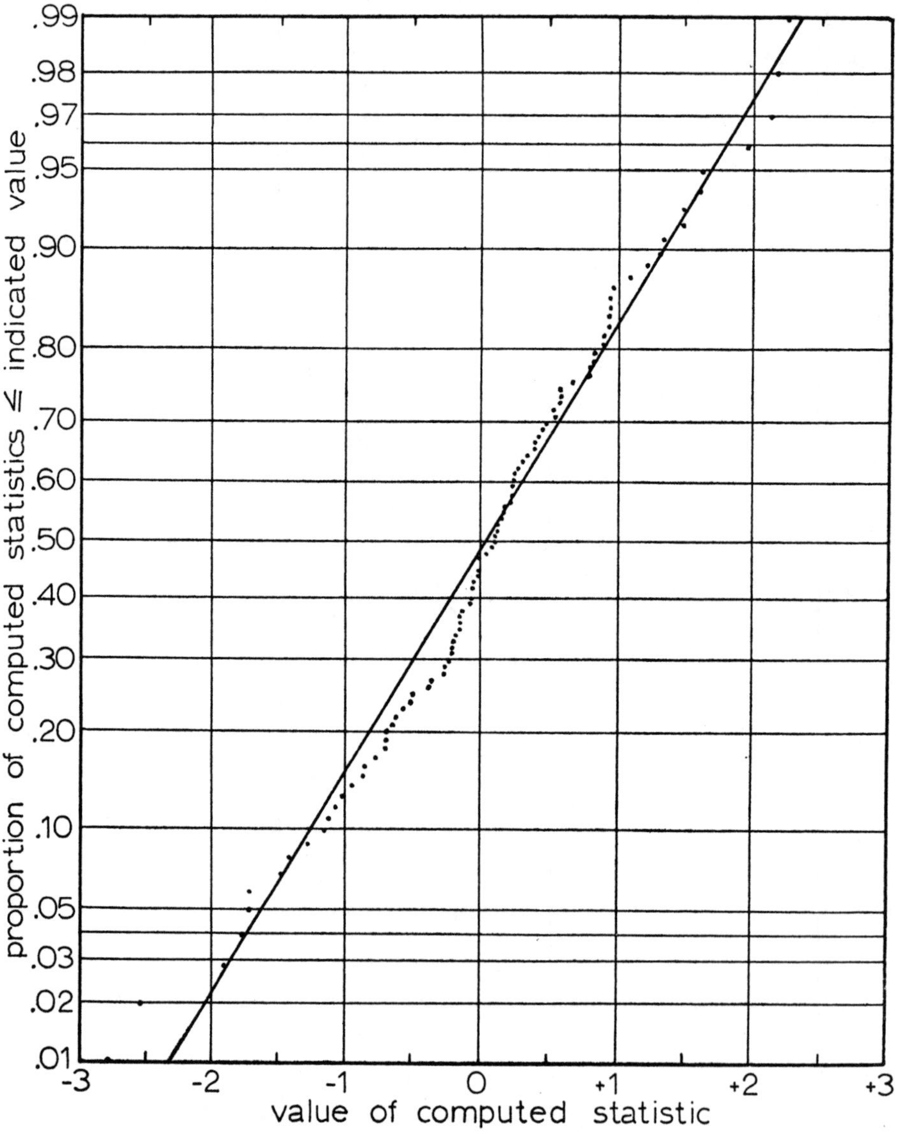

Straight line is unit-normal cumulative.
Unplotted maximum observed value is 2.61.

FIG. 3.8.7. Results of Sampling Experiment for (3.1.4). 50 Samples, Each of 200 Cross Classified Observations.

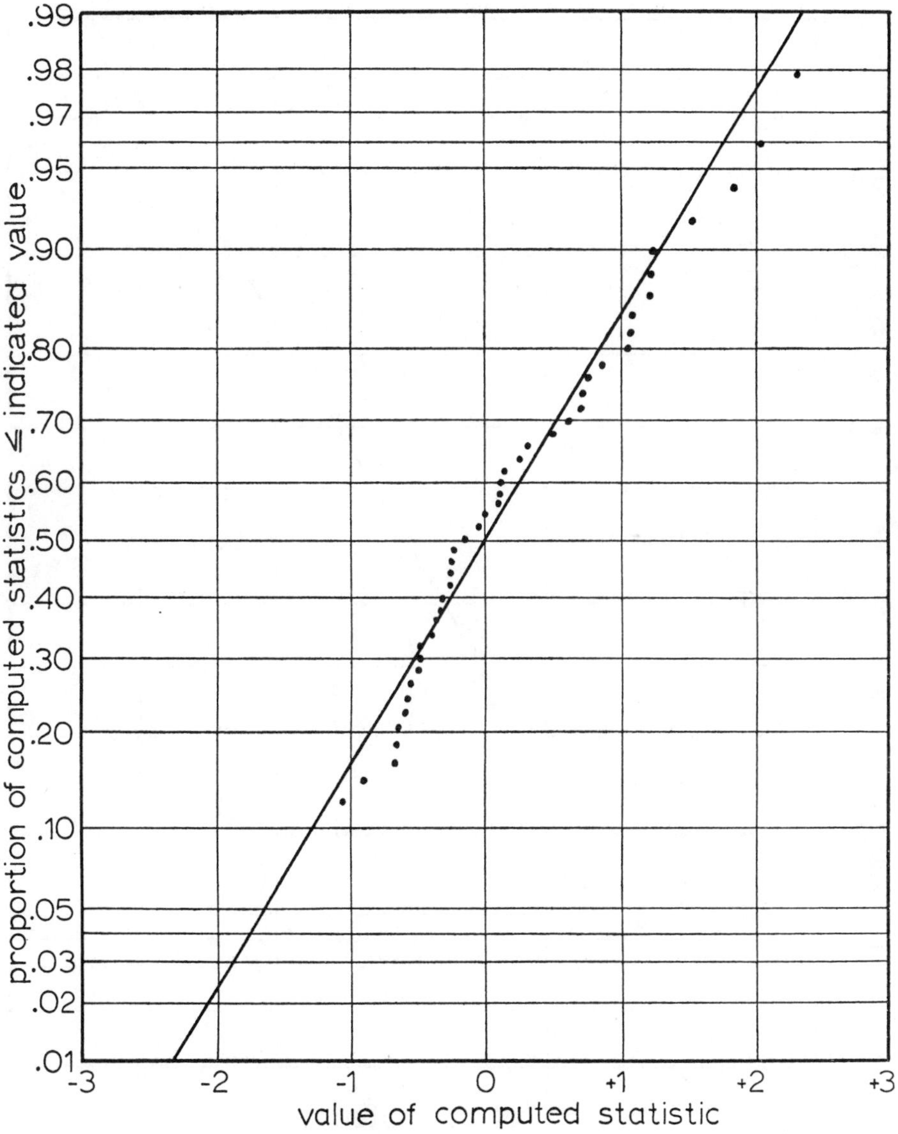

Straight line is unit-normal cumulative.
Unplotted maximum observed value is 2.88; five minimum values are $-\infty$.

FIG. 3.8.8. Results of Sampling Experiment for (3.3.4). 50 Samples, Each of 200 Cross Classified Observations.

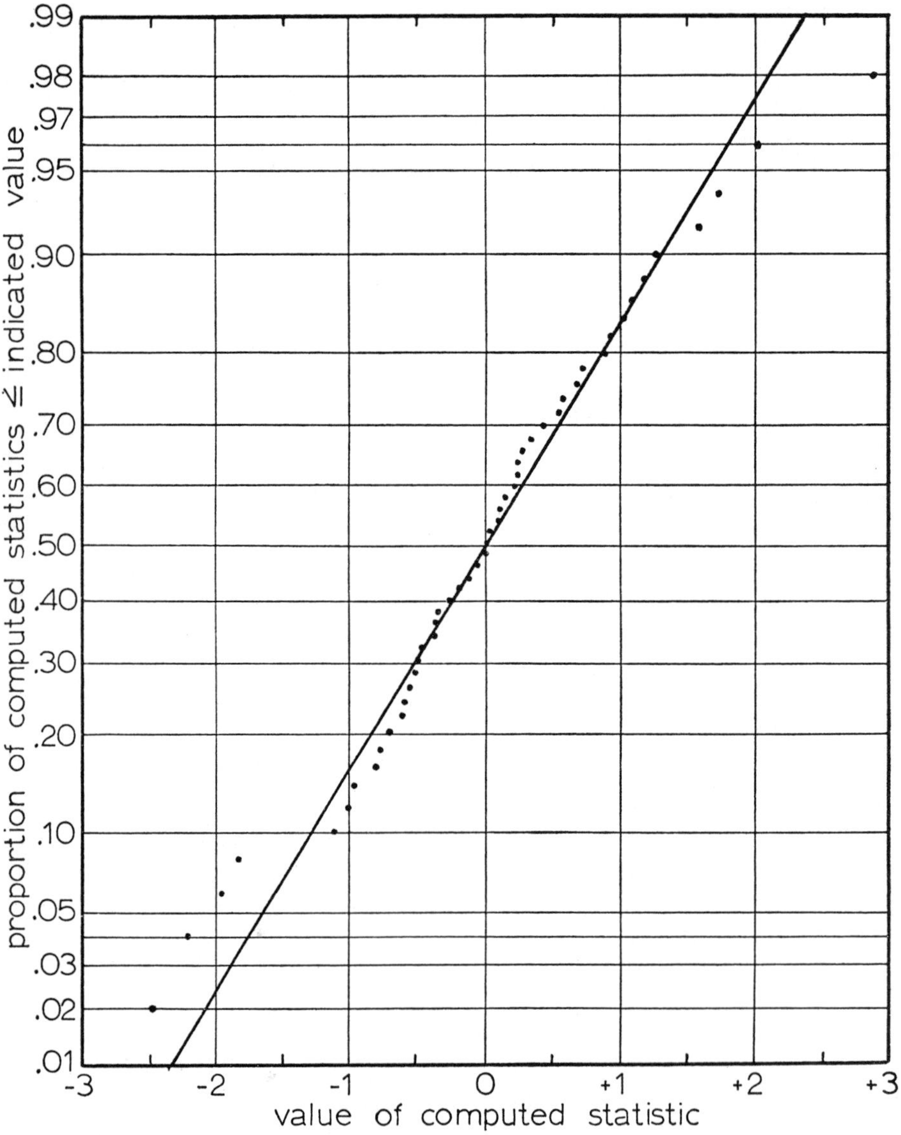

Straight line is unit-normal cumulative.
Unplotted maximum observed value is 3.01.

FIG. 3.8.9. Results of Sampling Experiment for $\sqrt{n}(G-\gamma)/\sqrt{(3.5.5)}$. 100 Samples, Each of 50 Cross Classified Observations.

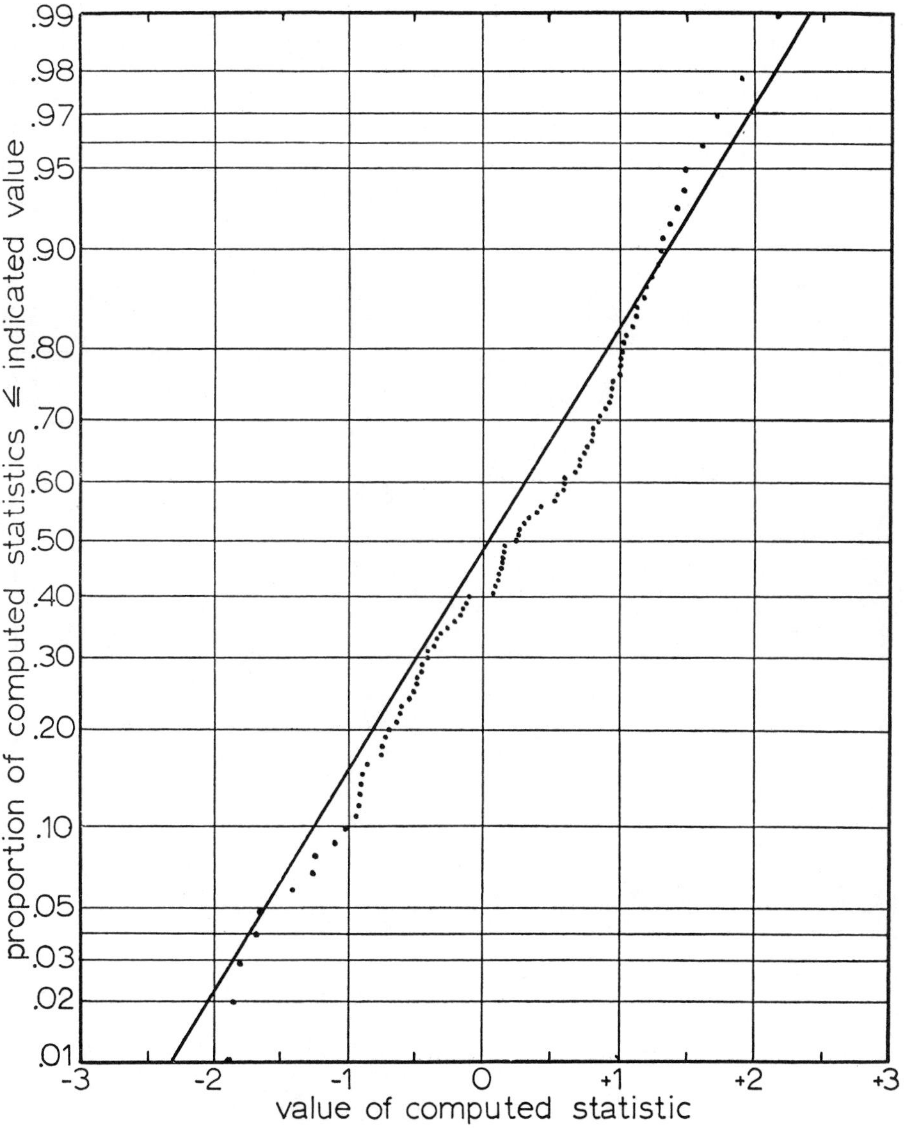

Straight line is unit-normal cumulative.
Unplotted maximum observed value is 2.23.

FIG. 3.8.10. Results of Sampling Experiment for $\sqrt{n}(G-\gamma)/\sqrt{(3.5.5)}$. 50 Samples, Each of 200 Cross Classified Observations.

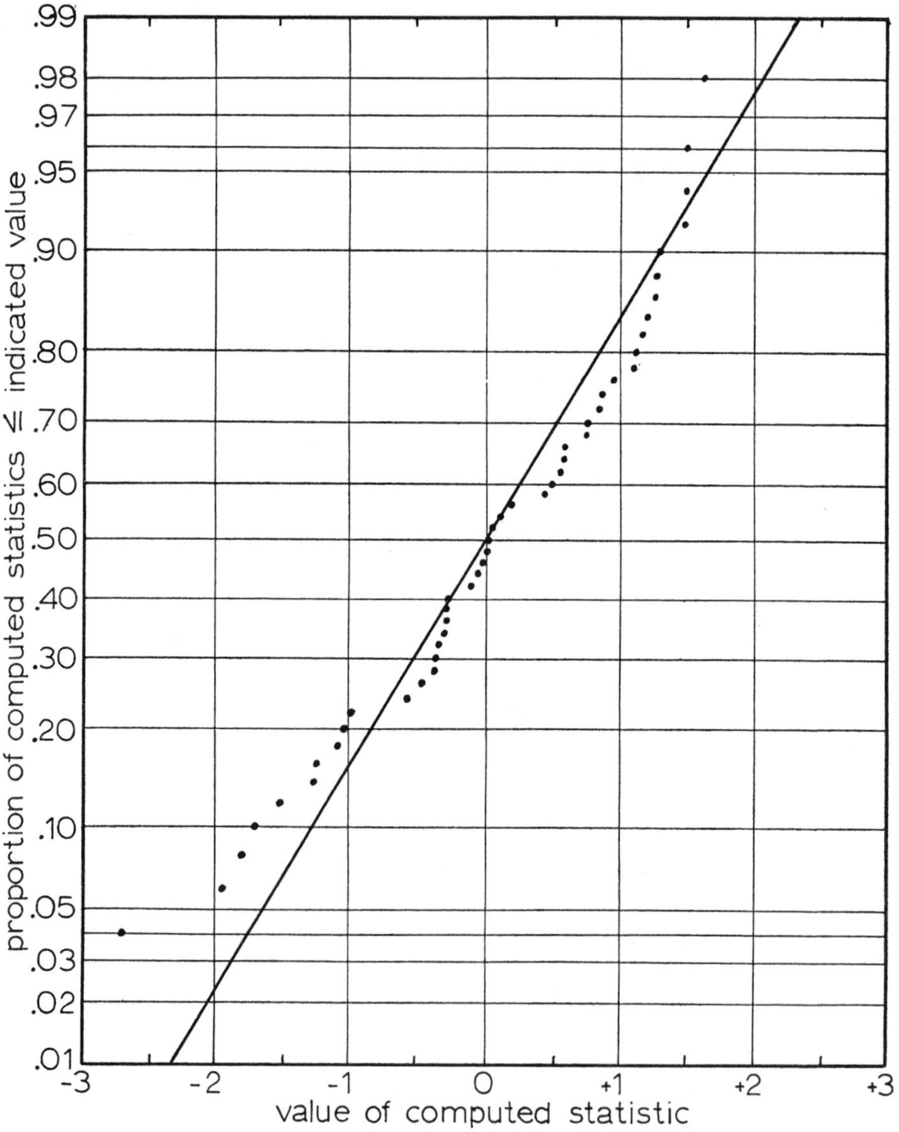

Straight line is unit-normal cumulative.
Unplotted maximum observed value is 1.75; minimum −3.14.

FIG. 3.8.11. Results of Sampling Experiment for $\sqrt{n}(G-\gamma)/\sqrt{(3.5.5)}$. 50 Samples, Each of 200 Cross Classified Observations.

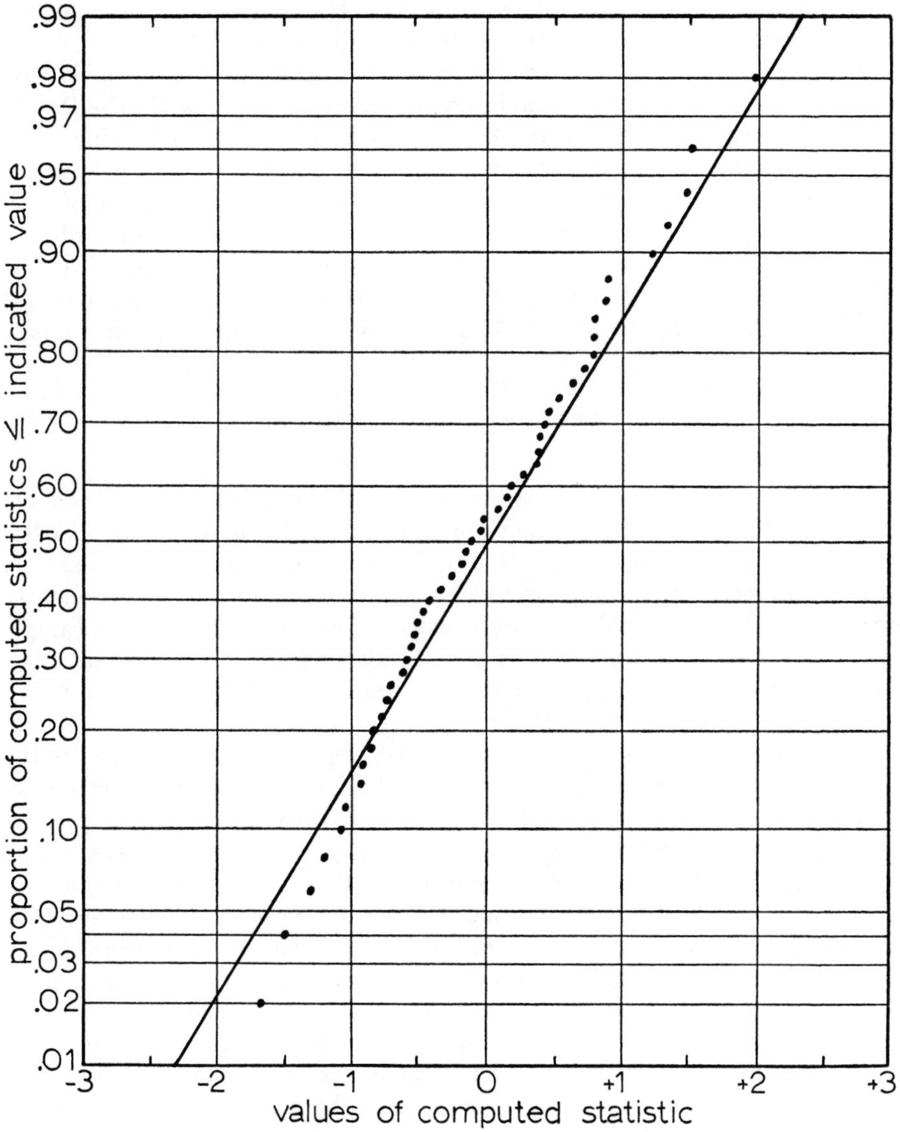

Straight line is unit-normal cumulative.
Unplotted maximum observed value is 2.83.

quencies from her sampling. The sample size, n, was 50 throughout, and 100 random samples were drawn from each of the 5×5 populations. Thus, for example, we see from Table 3.8.2 that there were eight 5×5 populations with γ values in the range [.20, .29].

The evidence of Table 3.8.2 suggests that the distribution of $\sqrt{n}\,(G-\gamma)/\sqrt{(3.5.5)}$ is reasonably well approximated by the unit normal distribution, for populations of the kind considered by Miss Rosenthal.

The distribution of (3.5.7), on the other hand, seems to be reasonably well approximated by the unit normal distribution, for populations of Miss Rosenthal's kind, only for values of $|\gamma|$ less than .5, and, even for such values of γ, there is a persistent tendency to have more frequent observations in the tails than the unit normal approximation predicts. This behavior is consistent with the presence in (3.5.7) of a random denominator. Of course, 50 can hardly be considered a large sample size, especially for 5×5 cross classifications! The distribution of (3.5.7) is more relevant to practice than that of $\sqrt{n}\,(G-\gamma)/\sqrt{(3.5.5)}$, and we plan to discuss (3.5.7) further in a subsequent publication.

4. MULTINOMIAL SAMPLING WITHIN EACH ROW (COLUMN) OF THE DOUBLE POLYTOMY

4.1. Preliminaries

Instead of sampling with replacement over all the cells of an $A \times B$ double polytomy, thus obtaining a sample point governed by a multinomial distribution, it may be necessary, desirable, or efficient to preassign and fix the sample size within each row (or, alternatively, column) and sample independently within each row (column) with replacement. Thus one will obtain a sample point governed by a product of α (or β) independent multinomial distributions.[13] For definiteness, suppose that the separate multinomials are within each row.

Since the sample sizes within rows are fixed, the N_{ab}'s are now subject to the α additional restrictions $\sum_b N_{ab} = n_{a\cdot}$, in addition to the over-all restriction $\sum_a \sum_b N_{ab} = n$. Within the ath row, the quantities $\sqrt{n_{a\cdot}}\,[(N_{ab}/n_{a\cdot}) - (\rho_{ab}/\rho_{a\cdot})]$ have zero means, variances $(\rho_{ab}/\rho_{a\cdot}) - (\rho_{ab}/\rho_{a\cdot})^2$, and covariances $-(\rho_{ab}/\rho_{a\cdot})(\rho_{ab'}/\rho_{a\cdot})$ for $b \neq b'$. As $n_{a\cdot} \to \infty$, the distribution of the quantities in question approaches that multivariate normal distribution with zero means and with the same variance-covariance structure. The N_{ab}'s in different rows are independent, both for finite samples and asymptotically. In considering asymptotic distributions for all rows together, we shall assume that, as $n \to \infty$, the ratios $n_{a\cdot}/n$ approach definite limits unequal to zero or one.

For this sampling procedure, and without ancillary knowledge, it is impossible to estimate the ρ_{ab}'s themselves, since the distribution of the sample point depends only upon the row-wise *conditional* probabilities $\rho_{ab}/\rho_{a\cdot}$. Hence, if we want to estimate or make tests on such measures as λ_b, λ, γ, etc. ,which themselves do not depend on the ρ_{ab}'s via the conditional probabilities $\rho_{ab}/\rho_{a\cdot}$, we must either assume the $\rho_{a\cdot}$'s known or perform a separate experiment to ob-

[13] One can also obtain such a product of distributions by starting with a multinomial distribution over the whole $\alpha \times \beta$ tableau, and then considering the conditional distribution *given* $N_{1\cdot} = n_{1\cdot}, \cdots, N_{\alpha\cdot} = n_{\alpha\cdot}$. This device and its extensions, are often used for the presentation of tests and the computation of so-called P-values in tests of independence with $\alpha \times \beta$ cross classifications.

tain estimates for them. If, however, we wish to estimate or make tests on a measure such as λ_b^* [8, Sec. 5.4], which is a function only of the ratios $\rho_{ab}/\rho_{a\cdot}$, then we may proceed without the assumption of knowledge of marginals and without performing a separate experiment. In the next three sub-sections we shall present examples of the above procedures.

4.2. The Index λ_b, With Marginal Row Probabilities Known

Perhaps the simplest case of the foregoing general discussion is that in which we are concerned with λ_b and in which

i) there is separate multinomial sampling in the rows,
ii) the row marginals, $\rho_{a\cdot}$, are known (we assume that they are positive),
iii) the maximum column marginal, $\rho_{\cdot m}$, is known, and
iv) sampling rates in the several rows are such that $n_{a\cdot}=n\rho_{a\cdot}$; that is, the sample sizes in the rows are proportional to the known row marginals.

Such a case might arise if the marginal probabilities are known, say, from census data, and if the sampling rates in the rows may be determined at our convenience.[14] We shall drop assumption (iii) shortly, but the development is simpler if we first suppose $\rho_{\cdot m}$ known.

A natural estimator for $\rho_{ab}/\rho_{a\cdot}$ is $N_{ab}/n_{a\cdot}=N_{ab}/(n\rho_{a\cdot})$. Hence N_{ab}/n is a natural estimator for ρ_{ab}. The corresponding estimator for λ_b is

$$L_b = \frac{\sum (N_{am}/n) - \rho_{\cdot m}}{1 - \rho_{\cdot m}}\,. \tag{4.2.1}$$

Strictly speaking, we should attach an added symbol to "L_b" in order to emphasize the fact that it is a different quantity from L_b of Section 3.1. We shall, however, refrain from doing this because of the already bothersome prolixity of notation.

The distribution of this new L_b depends only on the sum of the N_{am}'s. As before, we may, for asymptotic purposes, suppose that N_{am} is taken on in the same cell of the ath row in which ρ_{am} is taken. Hence, the quantities $\sqrt{n_{a\cdot}}$ $[(N_{am}/n_{a\cdot})-(\rho_{am}/\rho_{a\cdot})]=\sqrt{n}\ [(N_{am}/n)-\rho_{am}]/\sqrt{\rho_{a\cdot}}$ are asymptotically independent normal deviates with zero means and variances $(\rho_{am}/\rho_{a\cdot})[1-(\rho_{am}/\rho_{a\cdot})]$. (The expression for variance and the independence hold also for finite samples.) From this, we see that the quantities $\sqrt{n}[(N_{am}/n)-\rho_{am}]$ are asymptotically independent with zero means and variances $(\rho_{am}/\rho_{a\cdot})(\rho_{a\cdot}-\rho_{am})$. Hence, $\sqrt{n}\{\sum(N_{am}/n)-\sum \rho_{am}\}$ is asymptotically normal with zero mean and variance $\sum_a (\rho_{am}/\rho_{a\cdot})(\rho_{a\cdot}-\rho_{am})$. Since

$$\sqrt{n}(L_b - \lambda_b) = \sqrt{n}\{\textstyle\sum (N_{am}/n) - \sum \rho_{am}\}/(1 - \rho_{\cdot m}),$$

we have that $\sqrt{n}(L_b-\lambda_b)$ is asymptotically normal with zero mean and variance

$$\frac{1}{(1 - \rho_{\cdot m})^2}\sum_a \frac{\rho_{am}}{\rho_{a\cdot}}(\rho_{a\cdot} - \rho_{am}). \tag{4.2.2}$$

This is zero if and only if $\lambda_b=1$.

[14] Of course $n\rho_{a\cdot}$ will not in general be an integer, but for large n the difference will be unimportant, and asymptotically it makes no difference at all. We should also point out that setting $n_{a\cdot}=n\rho_{a\cdot}$ here is purely for convenience; we have not examined the question of whether or not this choice is good from the point of view of power.

Thus, if $\lambda_b \neq 1$, by the same argument as before, the following quantity is asymptotically unit-normal:

$$\sqrt{n}(L_b - \lambda_b) \frac{1 - \rho_{\cdot m}}{\sqrt{\sum \{\rho_{a\cdot}[N_{am}/n_{a\cdot}][1 - (N_{am}/n_{a\cdot})]\}}}, \tag{4.2.3a}$$

which may also be written as

$$n^{3/2}(L_b - \lambda_b) \frac{1 - \rho_{\cdot m}}{\sqrt{\sum \{N_{am}(n\rho_{a\cdot} - N_{am})/\rho_{a\cdot}\}}}. \tag{4.2.3b}$$

As an example of the method of using the normal approximation, suppose that $\alpha = \beta = 2$, $n = 30$, $n_{1\cdot} = 18$, $n_{2\cdot} = 12$, and that $\rho_{\cdot m} = 0.5$. (Note that $\rho_{\cdot m}$ need not be unique in this context.) Suppose further that the sample values turn out to be

$$N_{11} = 7 \qquad N_{12} = 11$$
$$N_{21} = 0 \qquad N_{22} = 12.$$

Then L_b is

$$\frac{1}{1 - .50}\left[\frac{(.6)(11)}{18} + \frac{(.4)(12)}{12} - .50\right] = .533.$$

The approximate 95% symmetrical confidence statement is that the following quantity lies between -1.96 and $+1.96$:

$$\frac{\sqrt{30}(.533 - \lambda_b)}{\sqrt{\left[\frac{(.6)(11)}{18}\frac{7}{18} + \frac{(.4)(12)}{12}\frac{0}{12}\right] \Big/ (1 - .50)^2}} = \frac{.533 - \lambda_b}{.138},$$

that is to say

$$.263 \leq \lambda_b \leq .803.$$

In order to obtain some idea of the adequacy of the normal approximation, the actual distribution of (4.2.3) was obtained in the following case:

$\alpha = \beta = 2$ $\qquad n_{1\cdot} = 18$ $\qquad n_{2\cdot} = 12$ $\qquad n = 30$

$\rho_{11} = .42$	$\rho_{12} = .18$	$\rho_{1\cdot} = .6$
$\rho_{21} = .08$	$\rho_{22} = .32$	$\rho_{2\cdot} = .4$
$\rho_{\cdot 1} = .50$	$\rho_{\cdot 2} = .50$	

$\lambda_b = .48$

and the results[15] are shown in Figure 4.2.1. The dots in Figure 4.2.1 appear at the "corners" of the discrete cumulative distribution graph of (4.2.3); for example, the probability that $(4.2.3) \leq 1.3717$ (to four places) is .86164 (to five places).

The approximation seems quite satisfactory for most statistical purposes in

[15] Because of the small numbers, it was feasible to find the exact distribution rather than resort to random sampling.

FIG. 4.2.1. Cumulative Distribution of (4.2.3).

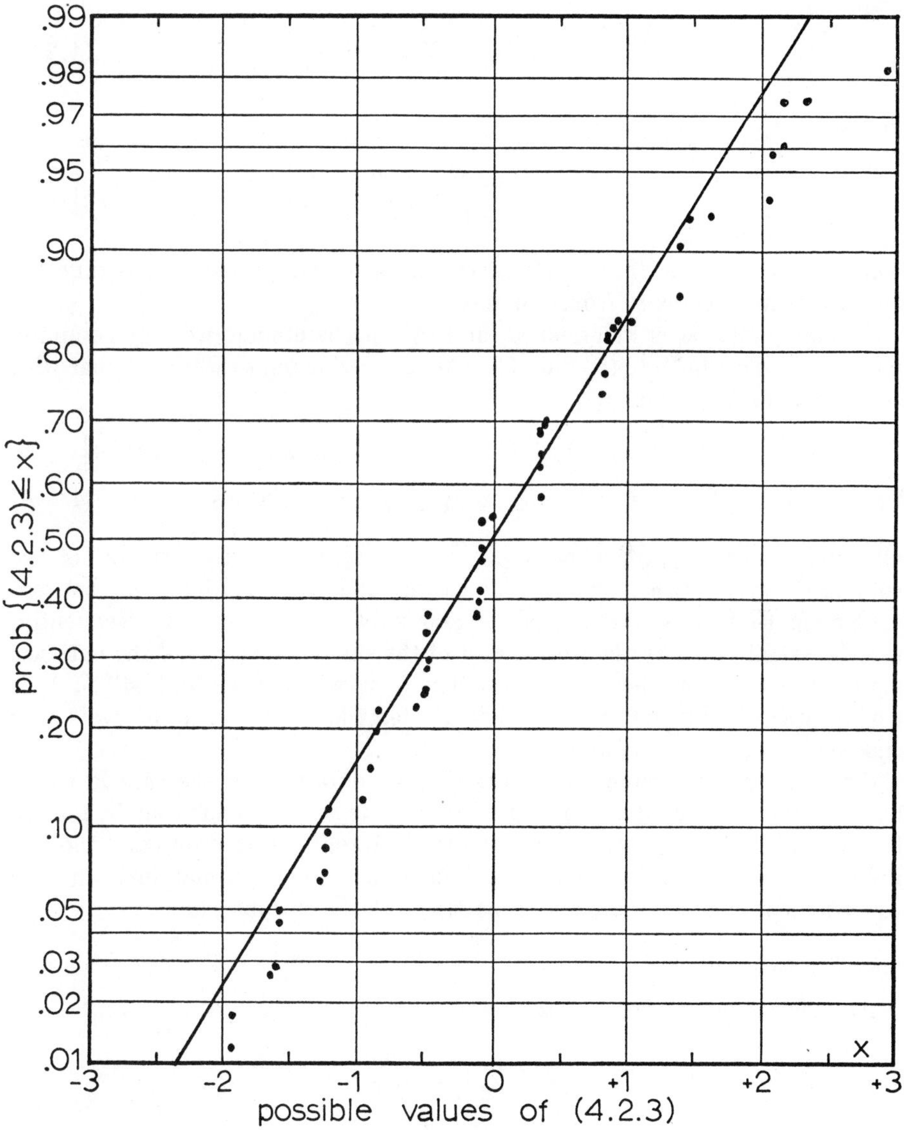

Straight line is unit-normal cumulative.

Three points omitted at left (lowest is at −2.63) and seven points omitted at right (highest is at 7.10).

this instance, and the sample size is by no means huge. Of course, this is only one case, and a rather special one at that.

It seems worthwhile to discuss briefly the results when assumption (iii) is dropped. In this case, $\rho_{\cdot m}$ is naturally estimated by the maximum likelihood estimator

$$\operatorname*{Max}_b (N_{\cdot b}/n) = N_{\cdot m}/n = R_{\cdot m}, \tag{4.2.4}$$

and λ_b by

$$L_b = \frac{\sum R_{am} - R_{\cdot m}}{1 - R_{\cdot m}}, \tag{4.2.5}$$

which is the *same* as (3.1.2). However, the probability structure is different because of independence from row-to-row.

Assuming that λ_b is well-defined, and that $\rho_{\cdot m}$ is unique, we may compute, much as in Section 3.1, that $\sqrt{n}(L_b - \lambda_b)$ is now asymptotically normal with zero mean and variance

$$(1 - \rho_{\cdot m})^2 \sum \rho_{am}[1 - (\rho_{am}/\rho_{a\cdot})] + (1 - \sum \rho_{am})^2 \sum \rho_{a(\cdot m)}[1 - (\rho_{a(\cdot m)}/\rho_{a\cdot})]$$
$$- 2(1 - \rho_{\cdot m})(1 - \sum \rho_{am})\{ \sum{}^{r} \rho_{am} - \sum (\rho_{am}\rho_{a(\cdot m)}/\rho_{a\cdot})\}, \tag{4.2.6}$$

all divided by $(1-\rho_{\cdot m})^4$, where $\rho_{a(\cdot m)}$ denotes that ρ_{ab} in the ath row of the column for which $\rho_{\cdot b}$ is maximum. The asymptotic variance just stated is zero (assuming that λ_b is well-defined) if and only if $\lambda_b = 0$ or 1. Consequently, $\sqrt{n}(L_b - \lambda_b)$ divided by the square root of the sample analogue of the variance stated above, under the assumption that λ_b is well-defined and $\neq 0$ or 1, is asymptotically unit normal. Approximate tests and confidence intervals may thus be found just as before.

Similar, but more complex, results may be obtained for the case in which only (i) and (ii) stated at the beginning of this section hold; that is, for the case in which there is separate multinomial sampling in the rows and the row marginals are known, but nothing is known of the column marginals and the row-wise sampling rates are *not* proportional to the row marginals.

4.3. The Index λ_b^*

An index of association, related to λ_b, is

$$\lambda_b^* = \frac{\sum \rho_{am}^* - \rho_{\cdot m}^*}{1 - \rho_{\cdot m}^*}, \tag{4.3.1}$$

where

$$\rho_{ab}^* = \rho_{ab}/(\alpha\rho_{a\cdot}), \quad \rho_{am}^* = \operatorname*{Max}_b \rho_{ab}^*, \quad \text{and} \quad \rho_{\cdot m}^* = \operatorname*{Max}_b \sum_a \rho_{ab}^*.$$

The use of this index was motivated in [8] by its independence of the row marginals $\rho_{a\cdot}$.

If we have separate multinomial sampling in the rows, a natural estimator of ρ^*_{ab} is $N_{ab}/(\alpha n_{a\cdot})$, and a natural estimator of λ^*_b is[16]

$$L_b^+ = \frac{\sum R^*_{am} - R^*_{\cdot m}}{1 - R^*_{\cdot m}}. \tag{4.3.2}$$

where

$$R^*_{ab} = R_{ab}/(\alpha R_{a\cdot}) = N_{ab}/(\alpha n_{a\cdot}), \quad R^*_{am} = \operatorname*{Max}_b R^*_{ab}, \quad \text{and} \quad R^*_{\cdot m} = \operatorname*{Max}_b \sum_a R^*_{ab}.$$

In order to approximate asymptotically the behavior of L_b^+ in this situation, we suppose that the $n_{a\cdot}$'s, or row-wise sampling rates, increase together so that $n_{a\cdot}/n \to \sigma_a$. With this hypothesis, we have from Section 4.1 that the quantities

$$\begin{aligned}\sqrt{n}(R^*_{ab} - \rho^*_{ab}) &= \sqrt{n}[(N_{ab}/n_{a\cdot}) - (\rho_{ab}/\rho_{a\cdot})]/\alpha \\ &\approx \sqrt{n_{a\cdot}}[(N_{ab}/n_{a\cdot}) - (\rho_{ab}/\rho_{a\cdot})]/(\alpha\sqrt{\sigma_a})\end{aligned}$$

are independent between rows, have zero means, and are asymptotically normal with variances

$$(\rho_{ab}/\rho_{a\cdot})[1 - (\rho_{ab}/\rho_{a\cdot})]/(\alpha^2\sigma_a) = \rho^*_{ab}(1 - \alpha\rho^*_{ab})/(\alpha\sigma_a).$$

As before, assume that ρ^*_{am} and $\rho^*_{\cdot m}$ are unique; whence, by a straightforward manipulation, $\sqrt{n}(L_b^+ - \lambda_b^*)$ is asymptotically normal with zero mean and variance

$$\begin{aligned}(1 - \rho^*_{\cdot m})^2 \sum [\rho^*_{am}(1 - \alpha\rho^*_{am})/(\alpha\sigma_a)] + (1 - \rho^*_{am})^2 \sum [\rho^*_{a(\cdot m)}(1 - \alpha\rho^*_{a(\cdot m)})/(\alpha\sigma_a)] \\ - 2(1 - \rho^*_{\cdot m})(1 - \rho^*_{am})\left\{\sum{}' \rho^*_{am}/(\alpha\sigma_a) - \sum \rho^*_{am}\rho^*_{a(\cdot m)}/\sigma_a\right\},\end{aligned}$$

all divided by $(1-\rho^*_{\cdot m})^4$. Here $\rho^*_{a(\cdot m)}$ is that ρ^*_{ab} in the ath row of the column for which $\sum_a \rho^*_{ab}$ is maximum; and $\sum'$ means a sum over those values of a for which $\rho^*_{am} = \rho^*_{a(\cdot m)}$. The asymptotic variance just stated is zero (assuming that λ^*_b is well defined) if and only if λ^*_b is 0 or 1.

Consequently, $\sqrt{n}(L_b^+ - \lambda_b^*)$, divided by the square root of the sample analogue of the variance stated above, under the assumption that λ^*_b is well defined and $\neq 0$ or 1, is asymptotically unit-normal. Approximate tests and confidence intervals may be found as before. If the sampling rate is the same in each row, i.e., if $\sigma_a = 1/\alpha$, then the asymptotic variance simplifies somewhat.

4.4. The Index τ_b, with Marginal Row Probabilities Known

In Section 9 of [8], we mentioned a measure of association based, not on optimal prediction, but on proportional prediction in a manner there explained. This measure is

$$\tau_b = \frac{\sum_a \sum_b \rho^2_{ab}/\rho_{a\cdot} - \sum_b \rho^2_{\cdot b}}{1 - \sum_b \rho^2_{\cdot b}}. \tag{4.4.1}$$

[16] The reason for using a plus sign as superscript, instead of an asterisk, is to emphasize that we are here dealing with separate multinomial sampling in the rows of the cross classification.

When conditions (i), (ii), and (iv), stated in Section 4.2, hold, a natural estimator for τ_b is

$$t_b = \frac{\sum_a \sum_b R_{ab}^2/r_{a\cdot} - \sum_b R_{\cdot b}^2}{1 - \sum_b R_{\cdot b}^2}, \tag{4.4.2}$$

where $R_{ab}=N_{ab}/n$, $r_{a\cdot}=n_{a\cdot}/n$, $R_{\cdot b}=N_{\cdot b}/n$. (The symbol t_b should not be confused with Stuart's t_c, referred to in Section 3.5; these measures of association are quite different.) It can be seen that $\sqrt{n}(t_b-\tau_b)$ is asymptotically normally distributed with zero mean and variance

* $$4\left\{\sum_{ab}\left[\frac{\rho_{ab}}{\rho_{a\cdot}}\left(1-\sum_b \rho_{\cdot b}^2\right) - \rho_{\cdot b}\left(1-\sum_{ab}\frac{\rho_{ab}^2}{\rho_{a\cdot}}\right)\right]^2 \rho_{ab} - \left[\sum_{ab}\frac{\rho_{ab}^2}{\rho_{a\cdot}} - \sum_b \rho_{\cdot b}^2\right]^2\right\}, \tag{4.4.3}$$

all divided by $(1-\sum_b \rho_{\cdot b}^2)^4$. Consequently, assuming that (4.4.1) is well defined and that (4.4.3) is different from zero, $\sqrt{n}(t_b-\tau_b)$ divided by the square root of the sample analogue of the variance (4.4.3) stated above is asymptotically unit-normal.

Similar, but more complex, results may be obtained in the situation where there is separate multinomial sampling in the rows and the row marginals are known, but the row-wise sampling rates are not necessarily proportional to the row marginals; i.e., the case where conditions (i) and (ii) of Section 4.2 hold true.

5. FURTHER REMARKS

The methods exemplified in this paper may be applied to other measures of association (or, more generally, to other sample measures), to other sampling procedures, to circumstances in which other kinds of outside knowledge exist, etc. One of our purposes has been to present these methods in a manner permitting their use by a wide class of research workers. (In some cases, different asymptotic methods may also be useful; for example, see Hoeffding [12].) In particular, one can obtain asymptotic approximations for the distributions of the traditional measures of association [8, Sec. 4] under other assumptions than that of independence. See [1], [27], [22], [21], [26], [15], [7]. Such approximations for the traditional measures have not been widely used. Perhaps one reason has been the almost obsessive interest in testing the null hypothesis of independence. Another possible reason is that the variance formulas obtained have been unwieldy and "too troublesome" to apply regularly in practice. See, for example, the editorial footnote on page 385 of [15].

We have not touched upon the question of power for the tests and confidence interval methods here discussed. The study of approximate power would be an appropriate next step.

6. REFERENCES

[1] Bulmer, M. G., "Confidence intervals for distance in the analysis of variance," *Biometrika*, 45 (1958), 360–9.

[2] Cramér, Harald, *Mathematical Methods of Statistics*. Princeton, New Jersey: Princeton University Press, 1946.

[3] Daniels, H. E., and Kendall, M. G., "The significance of rank correlations where parental correlation exists," *Biometrika*, 34 (1947), 198–208.

[4] Daniels, H. E., "Rank correlations and population models," *Journal of the Royal Statistical Society, Series B*, 12 (1950), 171–81.

[5] El-Bradry, M. A., and Stephan, F. F., "On adjusting sample tabulations to census counts," *Journal of the American Statistical Association*, 50 (1955), 738–62.

[6] Friedlander, D., "A technique for estimating a contingency table, given the marginal totals and some supplementary data," *Journal of the Royal Statistical Society, Series A*, 124 (1961), 412–20.

[7] Goodman, Leo A., "On methods for comparing contingency tables," *Journal of the Royal Statistical Society, Series A*, 126 (1963), 94–108.

[8] Goodman, Leo A., and Kruskal, William H., "Measures of association for cross classifications," *Journal of the American Statistical Association*, 49 (1954), 732–64.

[9] Goodman, Leo A., and Kruskal, William H., "Measures of association for cross classifications. II: Further discussion and references," *Journal of the American Statistical Association*, 54 (1959), 123–63.

[10] Greenwood, Robert E., and Glasgow, Mark O., "Distribution of maximum and minimum frequencies in a sample drawn from a multinomial distribution," *Annals of Mathematical Statistics*, 21 (1950), 416–24.

[11] Hoeffding (Höffding), Wassily, "On the distribution of the rank correlation coefficient τ when the variates are not independent," *Biometrika*, 34 (1947), 183–96.

[12] Hoeffding, Wassily, "A class of statistics with asymptotically normal distribution," *Annals of Mathematical Statistics*, 19 (1948), 293–325.

[13] Hoeffding, Wassily, and Robbins, Herbert, "The central limit theorem for dependent random variables," *Duke Mathematical Journal*, 15 (1948), 773–80.

[14] Hoeffding, Wassily, "An upper bound for the variance of Kendall's 'tau' and of related statistics," pp. 258–64 in *Contributions to Probability and Statistics. Essays in Honor of Harold Hotelling*, edited by Ingram Olkin and others, Stanford University Press, Stanford, 1960.

[15] Kondo, Tsutomu, "On the standard error of the mean square contingency," *Biometrika*, 21 (1929), 377–428.

[16] Kozelka, Robert M., "On some special order statistics from the multinomial distribution," Unpublished Ph.D. dissertation, Harvard University, 1952.

[17] Kozelka, Robert M., "Approximate upper percentage points for extreme values in multinomial sampling," *Annals of Mathematical Statistics*, 27 (1956), 507–12.

[18] Kruskal, William H., "Ordinal measures of association," *Journal of the American Statistical Association*, 53 (1958), 814–61.

[19] Mann, H. B., and Wald, A., "On stochastic limit and order relationships," *Annals of Mathematical Statistics*, 14 (1943), 217–26. (See also Wald, Abraham, *Selected Papers in Statistics and Probability*, McGraw-Hill Book Company, New York, 1955.)

[20] Noether, Gottfried E., "Two confidence intervals for the ratio of two probabilities and some measures of effectiveness," *Journal of the American Statistical Association*, 52 (1957), 36–45.

[21] Pearson, Karl, "On the probable error of a coefficient of mean square contingency," *Biometrika*, 10 (1915), 570–3.

[22] Sneyers, R., "Remarques à propos d'une généralisation de l'indice de similitude de M. Bouët," *Archiv für Meteorologie, Geophysik und Bioklimatologie, Series A: Meteorologie und Geophysik*, Band 11, 1. Heft, 1959, 126–35.

[23] Stuart, A., "The estimation and comparison of strengths of association in contingency tables," *Biometrika*, 40 (1953), 105–10.

[24] Sundrum, R. M., "Moments of the rank correlation coefficient τ in the general case," *Biometrika*, 40 (1953), 409–20.
[25] Weichselberger, K., "Über die Parameterschätzungen bei Kontingenztafeln, deren Randsummen vorgegeben sind," *Metrika*, 2 (1959), 100–30 and 198–229.
[26] Young, Andrew W., and Pearson, Karl, "On the probable error of a coefficient of contingency without approximation," *Biometrika*, 11 (1915), 215–30.
[27] Yule, G. Udny, "On the association of attributes in statistics with illustrations from the material from the childhood society, &c.," *Philosophical Transactions of the Royal Society of London, Series A*, 194 (1900), 257–319.

APPENDIX

A1. *Introduction.* The purpose of this appendix is to state some useful methods for deriving asymptotic distributions, to exemplify their use by outlining some of the derivations whose end-products are given in the preceding text, and to present other auxiliary material that seems inappropriate for inclusion in the text. An explicit statement of these methods may also be convenient for readers who wish to work out asymptotic distributions for other measures of association or for other sampling procedures than the ones discussed in this article.

A2. *A Basic Convergence Theorem.*

Theorem. If $\{X_n\}$ and $\{Y_n\}$ are two sequences of random variables, and if X is a random variable and y a constant, such that X_n converges in distribution to the distribution of X and Y_n converges in probability to y, then

X_n+Y_n converges in distribution to the distribution of $X+y$,
X_nY_n converges in distribution to the distribution of Xy,
X_n/Y_n converges in distribution to the distribution of X/y, (provided, in the last case, that $y\neq 0$).

(Convergence in distribution of X_n to X means that $\lim_{n\to\infty} \Pr\{X_n\leq x\} = \Pr\{X\leq x\}$ for every x at which $\Pr\{X\leq x\}$ is continuous. In the work of this paper the qualification may be neglected, for the limiting distributions are all normal, and hence continuous except for degenerate singular cases. For convergence in probability, see footnote 4.)

This result is stated, and an outline of its proof given, by Cramér [2, pp. 254–5]; it is a special case of earlier results by Mann and Wald (see [19], Theorem 5 and discussion on p. 224). The essential point is that if Z_n is a sequence of vector-valued random variables converging in distribution to the distribution of Z, and if f is a continuous vector-valued function (continuity may be weakened), then $f(Z_n)$ converges in distribution to the distribution of $f(Z)$.

This result has been applied in a number of places in the present text; e.g., in all places where a consistent estimator of the true asymptotic standard deviation of a statistic has been used, instead of the true asymptotic standard deviation itself, as the divisor of the difference between the statistic and its corresponding population value, without affecting the asymptotic distribution obtained.

A3. *The Delta Method Theorem.* We state this theorem for the case of two sequences of random variables, but its extension to more than two sequences is immediate. The proof of a more special form may be found in [2], pp. 366–7, and the proof of a somewhat more general form is in [13], pp. 777–8.

Theorem (Delta Method). Assume that W_n and V_n are two sequences of random variables ($n=1, 2, \cdots$) and that w and v are constants such that the pairs

$$\{\sqrt{n}(W_n - w), \sqrt{n}(V_n - v)\}$$

converge in distribution (bivariate sense) to the bivariate normal distribution with zero means, variances σ_{ww} and σ_{vv}, and covariance σ_{wv}.

Let $f(s, t)$ be a function with continuous first partial derivatives at (w, v).

Then $\sqrt{n}[f(W_n, V_n) - f(w, v)]$ is asymptotically normal, and in fact has the same asymptotic distribution as

$$\sqrt{n}\{a_w(W_n - w) + a_v(V_n - v)\},$$

where a_w, a_v are the partial derivatives of f with respect to its first and second arguments respectively, evaluated at (w, v). Thus the asymptotic distribution is the normal distribution with zero mean and variance

$$a_w^2\sigma_{ww} + 2a_wa_v\sigma_{wv} + a_v^2\sigma_{vv}.$$

Some of the results in this paper (e.g., the asymptotic variance (3.5.5)) were obtained with the aid of the delta method theorem. Some other results could be obtained either by applying this general theorem, or by more direct considerations (e.g., Section A5), since it is sometimes possible to arrange matters so that the function $f(s, t)$ is of the form a_ws+a_vt to begin with or so that a quantity simply related to $f(s, t)$ is of this linear form. We shall be frequently dealing with linear combinations of the $\sqrt{n}\ R_{ab}$'s, and we observe that the covariance structure of the $\sqrt{n}\ R_{ab}$'s is independent of n and is the same as the asymptotic covariance structure. Hence the variance of a linear combination of $\sqrt{n}\ R_{ab}$'s is exactly equal to its asymptotic variance.

A4. *Sample Maxima.* The fact that we may proceed, for asymptotic purposes, under the assumption that the R_{am}'s, $R_{\cdot m}$'s, etc., are taken on at the same columns and rows respectively as are the ρ_{am}'s, $\rho_{\cdot m}$'s, etc., is a consequence of the following.

Lemma. Let P_n ($n=1, 2, \cdots$) be a sequence of probability distributions, and let V be a chance event such that $\lim_{n\to\infty} P_n(V)$ exists; call it $P(V)$. Let W be a chance event such that $\lim_{n\to\infty} P_n(W)=1$. Then $\lim_{n\to\infty} P_n(V\cap W) = P(V)$. (The symbol "$\cap$" (read "and") means set-theoretic intersection.)

Proof: We note that

$$P_n(V) = P_n(V \cap W) + P_n(V \cap CW),$$

where CW is the complement of W. But $P_n(V\cap CW) \leq P_n(CW)$, which has the limit zero. This lemma is also a special case of the theorem of Section A2, obtained by replacing the probability of an event by the probability that its characteristic function is unity.

To apply this lemma, let P_n be the joint distribution of the random variables $\sqrt{n}(R_{ab}-\rho_{ab})$. Let W be the event defined by the requirement that the various sample maxima are taken on at the same columns and rows as the corresponding true maxima. V may be any chance event. For example, V might be the event that some function of the $\sqrt{n}(R_{ab}-\rho_{ab})$'s is $\leq$ a given constant. The hypotheses of the lemma are satisfied.

We are interested in the limit of $P_n(V)$. By applying the lemma, we see that this limit is equal to the limit of $P_n(V\cap W)$. This in turn is equal to the limit of $P_n(V'\cap W)$, where V' is like V except that R_{am} is replaced by that R_{ab} such that $\rho_{ab}=\rho_{am}$ (assuming uniqueness) and so on, since the event $V\cap W$ is exactly the same as the event $V'\cap W$. Finally, reapplying the lemma, the limit of $P_n(V'\cap W)$ is equal to the limit of $P_n(V')$, which is often easy to compute and which then gives us the desired limit of $P_n(V)$.

The following sections will illustrate applications of this lemma and the prior general theorems.

A5. *Asymptotic Behavior of L_b.* To examine the asymptotic behavior of L_b we write $\sqrt{n}(L_b-\lambda_b)$ as follows:

$$\sqrt{n}(L_b - \lambda_b) = \sqrt{n}\left(\frac{\sum R_{am} - R_{\cdot m}}{1 - R_{\cdot m}} - \frac{\sum \rho_{am} - \rho_{\cdot m}}{1 - \rho_{\cdot m}}\right)$$

$$= \sqrt{n}\, \frac{(\sum R_{am} - R_{\cdot m})(1 - \rho_{\cdot m}) - (\sum \rho_{am} - \rho_{\cdot m})(1 - R_{\cdot m})}{(1 - R_{\cdot m})(1 - \rho_{\cdot m})} \qquad \text{(A5.1)}$$

$$= \sqrt{n}\, \frac{[\sum (R_{am} - \rho_{am})](1 - \rho_{\cdot m}) - (R_{\cdot m} - \rho_{\cdot m})(1 - \sum \rho_{am})}{(1 - \rho_{\cdot m})^2} \cdot \frac{1 - \rho_{\cdot m}}{1 - R_{\cdot m}} .$$

The right-most factor, $(1-\rho_{\cdot m})/(1-R_{\cdot m})$, converges in probability to unity. Hence, if we can find the asymptotic distribution of what remains after omitting the right-most factor, we know, by Section A2, that it is the same as the asymptotic distribution *with* the right-most factor.

Since the ρ_{am}'s and $\rho_{\cdot m}$ are constants, our search then is effectively for the asymptotic distribution of $\sqrt{n}$ times a linear combination of the R_{am}'s and $R_{\cdot m}$; namely, that linear combination appearing in the numerator of the first fraction above. Let us call this numerator Δ. We already know, by the lemma presented in Section A4, that, for asymptotic purposes, R_{am} may be considered equal to R_{ab} for that value of b satisfying $\rho_{am}=\rho_{ab}$ and that similarly $R_{\cdot m}$ may be considered equal to $R_{\cdot b}$ for that value of b satisfying $\rho_{\cdot m}=\rho_{\cdot b}$.

Hence, by Section A3 and the facts just noted, it follows that $\sqrt{n}\Delta$ is asymptotically normal with mean zero and variance

$$\begin{aligned}(1 - \rho_{\cdot m})^2(\textstyle\sum \rho_{am})(1 - \sum \rho_{am})& \\ + (1 - \textstyle\sum \rho_{am})^2\rho_{\cdot m}(1 - \rho_{\cdot m})& \qquad \text{(A5.2)}\\ - 2(1 - \rho_{\cdot m})(1 - \textstyle\sum \rho_{am})[\sum' \rho_{am} - \rho_{\cdot m} \sum \rho_{am}].&\end{aligned}$$

For the reader's convenience, we insert here the argument leading to the above expression; similar arguments can be presented to obtain derivations of the asymptotic variances, presented in the text, for some of the other statistics discussed.

The first line of (A5.2) is the variance of the first term of $\sqrt{n}\Delta$; viz., $\sqrt{n}(1-\rho_{\cdot m})\sum(R_{am}-\rho_{am})$. The constant term, $(1-\rho_{\cdot m})$, is squared, and, since $\sum R_{am}$ and $1-\sum R_{am}$ may be considered as the proportions of successes and failures in n independent Bernouilli trials with constant probabilities $\sum\rho_{am}$ and $1-\sum\rho_{am}$, the variance of $\sqrt{n}\sum(R_{am}-\rho_{am})$ or of $\sqrt{n}\sum R_{am}$ is simply $(\sum\rho_{am})(1-\sum\rho_{am})$.

The second line of (A5.2) is the variance of the second term of $\sqrt{n}\Delta$; viz., $\sqrt{n}(1-\sum\rho_{am})(R_{\cdot m}-\rho_{\cdot m})$. It may be written down in the same manner as the first line.

The third line of (A5.2) is minus twice the covariance of the two terms of $\sqrt{n}\Delta$. The constants afford no difficulty, but the quantity in square brackets, the covariance of $\sqrt{n}\sum R_{am}$ and $\sqrt{n}R_{\cdot m}$, may be troublesome to check. Note that, in general, if $R_1, R_2, \cdots, R_k$ are multinomial proportions corresponding to the probabilities $\rho_1, \rho_2, \cdots, \rho_k$, then

$$\begin{aligned}\operatorname{Cov}(R_1+R_2, R_1+R_3) &= \operatorname{Var} R_1 + \operatorname{Cov}(R_1,R_2) + \operatorname{Cov}(R_1,R_3) + \operatorname{Cov}(R_2,R_3)\\ &= n^{-1}\{\rho_1(1-\rho_1) - \rho_1\rho_2 - \rho_1\rho_3 - \rho_2\rho_3\}\\ &= n^{-1}\{\rho_1 - (\rho_1+\rho_2)(\rho_1+\rho_3)\}.\end{aligned}$$

To apply this, take R_1 as the sum of those R_{am}'s that also appear as summands of $R_{\cdot m}$, R_2 as $\sum R_{am}-R_1$, and R_3 as $R_{\cdot m}-R_1$. We can therefore write down the third line above immediately.

Simplifying the above expression for the variance of the asymptotic distribution of $\sqrt{n}\Delta$, we obtain

$$(1-\rho_{\cdot m})(1-\sum\rho_{am})(\sum\rho_{am} + \rho_{\cdot m} - 2\sum{}^{r}\rho_{am}). \tag{A5.3}$$

Note that the last factor is just the sum of the ρ_{am}'s not summands in $\rho_{\cdot m}$, plus those summands in $\rho_{\cdot m}$ that are not ρ_{am}'s. (A5.3) is zero if and only if $\lambda_b=0$.

Hence, as $n\to\infty$, the distribution of $\sqrt{n}(L_b-\lambda_b)$ will approach the normal distribution with mean zero and variance (3.1.3). This variance is zero if and only if $\lambda_b=0$ or 1. It is indeterminate if $\rho_{\cdot m}=1$, but in this case λ_b itself is indeterminate. Finally, if $\lambda_b\neq 0$ or 1, it follows from Section A2 that (3.1.4) is asymptotically unit normal.

We note that (3.1.3) could also be obtained by direct application of the delta method to the first line of (A5.1). We have that

$$\frac{\partial}{\partial R_{\cdot m}}\left(\frac{\sum R_{am}-R_{\cdot m}}{1-R_{\cdot m}}\right) = -\frac{1-\sum R_{am}}{(1-R_{\cdot m})^2},$$

and

$$\frac{\partial}{\partial \sum R_{am}}\left(\frac{\sum R_{am}-R_{\cdot m}}{1-R_{\cdot m}}\right) = \frac{1}{1-R_{\cdot m}}.$$

Evaluating these derivatives for the population values, we obtain $-(1-\sum\rho_{am})/(1-\rho_{\cdot m})^2$ and $1/(1-\rho_{\cdot m})$, respectively. If we then computed the quadratic form required by the delta method, and used the appropriate variances and covariances given a few paragraphs back, we would emerge with (3.1.3). These two paths to the same goal serve as mutual checks.

A6. *Asymptotic Behavior of L.* Just as in Section A5, we write $\sqrt{n}(L-\lambda)$ as one fraction and change the denominator to a function of the ρ's while at the same time multiplying by a quantity approaching unity in probability. Neglecting this last quantity, as we may by Section A2, we deal with

$$\frac{\sqrt{n}}{(2-\rho_{\cdot m}-\rho_{m\cdot})^2}\{(2-\rho_{\cdot m}-\rho_{m\cdot})(\sum R_{am}+\sum R_{mb}) + (\sum \rho_{am}+\sum \rho_{mb}-2)(R_{\cdot m}+R_{m\cdot}) + 2(\rho_{\cdot m}+\rho_{m\cdot}-\sum \rho_{am}-\sum \rho_{mb})\}. \tag{A6.1}$$

This quantity is asymptotically normal with zero mean and with a variance that is equal to the following divided by $(2-\rho_{\cdot m}-\rho_{m\cdot})^4$:

$$\begin{aligned}&(2-\rho_{\cdot m}-\rho_{m\cdot})^2[\sum \rho_{am}(1-\sum \rho_{am})+\sum \rho_{mb}(1-\sum \rho_{mb}) \\ &\qquad + 2\sum{}^{*} \rho_{am}-2(\sum \rho_{am})(\sum \rho_{mb})] \\ &+(2-\sum \rho_{am}-\sum \rho_{mb})^2[\rho_{\cdot m}(1-\rho_{\cdot m})+\rho_{m\cdot}(1-\rho_{m\cdot})+2\rho_{**}-2\rho_{m\cdot}\rho_{\cdot m}] \\ &-2(2-\rho_{\cdot m}-\rho_{m\cdot})(2-\sum \rho_{am}-\sum \rho_{mb})[\sum{}^{r} \rho_{am}-\rho_{\cdot m}\sum \rho_{am}+\sum{}^{c} \rho_{mb} \\ &\qquad -\rho_{m\cdot}\sum \rho_{mb}+\rho_{*m}-\rho_{m\cdot}\sum \rho_{am}+\rho_{m*}-\rho_{\cdot m}\sum \rho_{mb}],\end{aligned} \tag{A6.2}$$

where the notation involving asterisks is defined in Section 3.4 in terms of sample R's. Simplifying (A6.2) we obtain the quantity (A6.3),[17]

$$\begin{aligned}&(2-\rho_{\cdot m}-\rho_{m\cdot})^2[(\sum \rho_{am}+\sum \rho_{mb})(1-\sum \rho_{am}-\sum \rho_{mb})+2\sum{}^{*} 2\rho_{am}] \\ &+(2-\sum \rho_{am}-\sum \rho_{mb})^2[(\rho_{\cdot m}+\rho_{m\cdot})(1-\rho_{\cdot m}-\rho_{m\cdot})+2\rho_{**}] \\ &-2(2-\rho_{\cdot m}-\rho_{m\cdot})(2-\sum \rho_{am}-\sum \rho_{mb})[\sum{}^{r} \rho_{am}+\sum{}^{c} \rho_{mb}+\rho_{*m} \\ &\qquad +\rho_{m*}-(\rho_{\cdot m}+\rho_{m\cdot})(\sum \rho_{am}+\sum \rho_{mb})].\end{aligned} \tag{A6.3}$$

If, for simplicity, we let

$$\begin{aligned}\Upsilon_{\cdot} &= \rho_{\cdot m}+\rho_{m\cdot}, \\ \Upsilon_{\Sigma} &= \sum \rho_{am}+\sum \rho_{mb}, \\ \Upsilon_{*} &= \sum{}^{r} \rho_{am}+\sum{}^{c} \rho_{mb}+\rho_{*m}+\rho_{m*},\end{aligned}$$

then (A6.3) is equal to both the following quantities:

$$(2-\Upsilon_{\cdot})^2(\Upsilon_{\Sigma}+2\sum{}^{*} \rho_{am})+(2-\Upsilon_{\Sigma})^2(\Upsilon_{\cdot}+2\rho_{**})-4(\Upsilon_{\Sigma}-\Upsilon_{\cdot})^2 - 2(2-\Upsilon_{\cdot})(2-\Upsilon_{\Sigma})\Upsilon_{*} \tag{A6.4a}$$

$$(2-\Upsilon_{\cdot})(2-\Upsilon_{\Sigma})(\Upsilon_{\cdot}+\Upsilon_{\Sigma}+4-2\Upsilon_{*})-2(2-\Upsilon_{\cdot})^2(1-\sum{}^{*} \rho_{am}) -2(2-\Upsilon_{\Sigma})^2(1-\rho_{**}). \tag{A6.4b}$$

---

[17] Actually we could write (A6.3) directly by the use of general formulas such as that of Section A5. For example it is easy to check that

$$\mathrm{Var}[(R_1+R_2)+(R_1+R_3)] = n^{-1}[(2\rho_1+\rho_2+\rho_3)(1-2\rho_1-\rho_2-\rho_3)].$$

The variance of the asymptotic normal distribution for $\sqrt{n}(L-\lambda)$ is either of the above divided by $(2-\Upsilon.)^4$.

It follows from Section A2 that, provided (A6.4) is not zero, (3.4.3) is asymptotically unit normal.

Finally, we show that (A6.4) is zero if and only if $\lambda=0$ or 1. We are concerned with the variance of

$$(2 - \Upsilon.)U_\Sigma + (\Upsilon_\Sigma - 2)U., \qquad \text{(A6.5)}$$

the random quantity in (A6.1). To say that this has zero variance is to say that it is constant; we consider the various possible cases.

(A): $2-\Upsilon.=0$. This says that $\rho_{\cdot m}=\rho_{m\cdot}=1$, or that there is only one non-zero ρ_{ab}, and it equals unity. This degenerate case, in which λ is not even defined, we have precluded by assumption.

(B): $2-\Upsilon_\Sigma=0$. This says that $\sum\rho_{am}=\sum\rho_{mb}=1$, or that each row and column has at most one non-zero ρ_{ab}. Then $\lambda=1$, $L=1$ always, and $U_\Sigma=2$ always. Thus, in this special case, $\lambda=1$ and the variance of (A6.5) is zero.

(C): $2-\Upsilon.\neq0$, $2-\Upsilon_\Sigma\neq0$. If neither coefficient of (A6.5) is zero, at least one row or column has two or more cells with non-zero ρ_{ab}'s. Without loss of generality, suppose that ρ_{11} and ρ_{12} are both non-zero. Then there is positive probability for each of the following $n+1$ sample points: $R_{11}=k/n$, $R_{12}=(n-k)/n$ $(k=0, 1, \cdots, n)$. If $k\leq n/2$, $U_\Sigma=U.=1+[(n-k)/n]$; while if $k\geq n/2$, $U_\Sigma=U.=1+[k/n]$. Hence, for such sample points,

$$\text{(A6.5)} = (\Upsilon_\Sigma - \Upsilon.)[1 + \operatorname{Max}(k, n-k)/n], \qquad (k = 0, 1, \cdots, n).$$

If (A6.5) is constant and $n\geq2$, it follows that $\Upsilon_\Sigma=\Upsilon.$, or $\lambda=0$.

Conversely, it is easy to see that if $\lambda=0$, all the ρ_{am}'s must appear in the same column, and all the ρ_{mb}'s in the same row. Hence $\Upsilon.=\Upsilon_\Sigma$, $\Upsilon_*=\Upsilon.+2\rho_{**}$, and $\Sigma^*\rho_{am}=\rho_{**}=\rho_{*m}=\rho_{m*}$. Hence, substituting in (A6.4), we obtain zero variance.

A7. *Asymptotic Behavior of G.* As before, we simplify

$$\sqrt{n}(G - \gamma) = \sqrt{n}\left[\frac{P_s - P_d}{1 - P_t} - \frac{\Pi_s - \Pi_d}{1 - \Pi_t}\right]$$

by writing it as a single fraction, and then replacing the denominator by the quantity to which it converges in probability. This leaves us with

$$\frac{2}{(1 - \Pi_t)^2}\sqrt{n}[P_s\Pi_d - P_d\Pi_s] \qquad \text{(A7.1)}$$

as the quantity whose asymptotic distribution is desired. We assume that $\Pi_t<1$. By use of the delta method (Section A3), we see that $\sqrt{n}[P_s\Pi_d-P_d\Pi_s]$ is asymptotically normal with zero mean and with a variance computed in the following manner. The random variable P_s, considered as a function of the R_{ab}'s, may be partially differentiated with respect to R_{ab} as follows:

$$\frac{\partial P_s}{\partial R_{\mathbf{ab}}} = 2 \sum_a \sum_b \sum_{a'>a} \sum_{b'>b} \frac{\partial}{\partial R_{\mathbf{ab}}} R_{ab} R_{a'b'}.$$

Unless $(a, b) = (\mathbf{a}, \mathbf{b})$ or $(a', b') = (\mathbf{a}, \mathbf{b})$, the summand is zero. (Both equalities cannot hold simultaneously.) Hence, we obtain

$$\frac{\partial P_s}{\partial R_{\mathbf{ab}}} = 2 \sum_{a'>\mathbf{a}} \sum_{b'>\mathbf{b}} R_{a'b'} + 2 \sum_{a'<\mathbf{a}} \sum_{b'<\mathbf{b}} R_{a'b'}. \tag{A7.2a}$$

Similarly,

$$\frac{\partial P_d}{\partial R_{\mathbf{ab}}} = 2 \sum_{a'>\mathbf{a}} \sum_{b'<\mathbf{b}} R_{a'b'} + 2 \sum_{a'<\mathbf{a}} \sum_{b'>\mathbf{b}} R_{a'b'}. \tag{A7.2b}$$

Evaluating these at $R_{a'b'} = \rho_{a'b'}$, and changing $(\mathbf{a}, \mathbf{b})$ to (a, b) we obtain the differential coefficients

$$\begin{aligned} 2\mathcal{R}_{ab}^{(s)} &= \left[\frac{\partial P_s}{\partial R_{ab}}\right]_{R_{a'b'}=\rho_{a'b'}} = 2 \sum_{a'>a} \sum_{b'>b} \rho_{a'b'} + 2 \sum_{a'<a} \sum_{b'<b} \rho_{a'b'}, \\ 2\mathcal{R}_{ab}^{(d)} &= \left[\frac{\partial P_d}{\partial R_{ab}}\right]_{R_{a'b'}=\rho_{a'b'}} = 2 \sum_{a'>a} \sum_{b'<b} \rho_{a'b'} + 2 \sum_{a'<a} \sum_{b'>b} \rho_{a'b'}. \end{aligned} \tag{A7.3}$$

Essentially these quantities appear in (3.5.6).

Hence the derivative of $P_s\Pi_d - P_d\Pi_s$ with respect to R_{ab}, evaluated for the true $\rho_{a'b'}$'s, is

$$2[\Pi_d \mathcal{R}_{ab}^{(s)} - \Pi_s \mathcal{R}_{ab}^{(d)}],$$

and the asymptotic variance of $\sqrt{n}[P_s\Pi_d - P_d\Pi_s]$ is the same as the asymptotic variance of

$$\sum_a \sum_b 2[\Pi_d \mathcal{R}_{ab}^{(s)} - \Pi_s \mathcal{R}_{ab}^{(d)}]\sqrt{n}(R_{ab} - \rho_{ab}). \tag{A7.4}$$

But this quantity is

$$4 \sum_a \sum_b \sum_{a'} \sum_{b'} [\Pi_d \mathcal{R}_{ab}^{(s)} - \Pi_s \mathcal{R}_{ab}^{(d)}][\Pi_d \mathcal{R}_{a'b'}^{(s)} - \Pi_s \mathcal{R}_{a'b'}^{(d)}][\delta_{ab,a'b'}\rho_{ab} - \rho_{ab}\rho_{a'b'}], \tag{A7.5}$$

where $\delta_{ab,a'b'} = 0$ unless $(a, b) = (a', b')$, in which case it is unity. Now consider the eight terms of the above sum, obtained by multiplying out. The four terms that involve $\delta_{ab,a'b'}$ are

$$4 \sum_a \sum_b \Pi_d^2 \mathcal{R}_{ab}^{(s)2} \rho_{ab} = 4\Pi_d^2 \Pi_{ss},$$

$$-4 \sum_a \sum_b \Pi_s \Pi_d \mathcal{R}_{ab}^{(s)} \mathcal{R}_{ab}^{(d)} \rho_{ab} = -4\,\Pi_s \Pi_d \Pi_{sd} \text{ (two of these), and}$$

$$4 \sum_a \sum_b \Pi_s^2 \mathcal{R}_{ab}^{(d)2} \rho_{ab} = 4\Pi_s^2 \Pi_{dd};$$

whereas the other terms sum to zero, for they are equal to

$$4\left\{ \sum_a \sum_b \left[\Pi_d \mathcal{R}_{ab}^{(s)} - \Pi_s \mathcal{R}_{ab}^{(d)}\right]\rho_{ab}\right\}^2$$

and

$$\Pi_d \sum_a \sum_b \rho_{ab} \mathcal{R}_{ab}^{(s)} = \Pi_d \Pi_s = \Pi_s \sum_a \sum_b \rho_{ab} \mathcal{R}_{ab}^{(d)}.$$

Hence the asymptotic variance of $\sqrt{n}[P_s\Pi_d - P_d\Pi_s]$ is

$$4[\Pi_d^2\Pi_{ss} - 2\Pi_s\Pi_d\Pi_{sd} + \Pi_s^2\Pi_{dd}].$$

Finally, we must multiply this by the square of the constant coefficient in (A7.1) to obtain (3.5.5), the asymptotic variance of $\sqrt{n}(G-\gamma)$.

It is interesting to note that (3.5.5) reduces, in the continuous case, to

$$16\{\Pi_s^2\Pi_d - \Pi_s^2\Pi_{sd} - 2\Pi_s\Pi_d\Pi_{sd} + \Pi_d^2\Pi_s - \Pi_d^2\Pi_{sd}\} = 16\{\Pi_{ss} - \Pi_s^2\}, \quad \text{(A7.6)}$$

since, in this case, $\Pi_t = 0$, $\Pi_s = \Pi_{ss} + \Pi_{sd}$, $\Pi_d = \Pi_{dd} + \Pi_{sd}$, $\Pi_s + \Pi_d = 1$. Formula (A7.6) is, in fact, the asymptotic variance for $\sqrt{n}(t-\tau)$, where t is Kendall's t (see [11], [24], [18]). This is natural, since G, in the continuous case, is [18] essentially Kendall's t.

We now turn to the bound on the asymptotic variance stated in Section 3.5. This quantity is also a bound on the actual (small-sample) variance of $\sqrt{n}$ $(G-\gamma)$, at least for even-sized samples [4]. We shall prove here that (3.5.5) is less than or equal to the bound (3.5.9), $2(1-\gamma^2)/(1-\Pi_t)$; i.e., that

$$\{\Pi_s^2\Pi_{dd} - 2\Pi_s\Pi_d\Pi_{sd} + \Pi_d^2\Pi_{ss}\} \leq (1 - \Pi_t)^3(1 - \gamma^2)/8. \quad \text{(A7.7)}$$

Proof: Let

$$V_{ij} = \begin{cases} -\Pi_d & \text{if observations } i \text{ and } j \text{ are fully concordant.} \\ \Pi_s & \text{if observations } i \text{ and } j \text{ are fully discordant.} \\ 0 & \text{otherwise.} \end{cases}$$

Then

$$\begin{aligned} E\{V_{ij}\} &= -\Pi_d\Pi_s + \Pi_s\Pi_d = 0, \\ \mathrm{Var}\{V_{ij}\} &= \Pi_d^2\Pi_s + \Pi_s^2\Pi_d = \Pi_s\Pi_d(\Pi_d + \Pi_s) = \Pi_s\Pi_d(1 - \Pi_t), \\ \mathrm{Cov}\{V_{ij}, V_{ik}\} &= \Pi_d^2\Pi_{ss} - 2\Pi_s\Pi_d\Pi_{sd} + \Pi_s^2\Pi_{dd}, \end{aligned}$$

for $j \neq k$. Since

$$\begin{aligned} &\mathrm{Var}\{2(V_{12} + V_{34}) - (V_{13} + V_{14} + V_{23} + V_{24})\} \\ &\quad = 12\,\mathrm{Var}\{V_{12}\} - 24\,\mathrm{Cov}\{V_{12}, V_{13}\} \geq 0, \end{aligned}$$

we have that

$$\Pi_s\Pi_d(1 - \Pi_t) \geq 2\{\Pi_d^2\Pi_{ss} - 2\Pi_s\Pi_d\Pi_{sd} + \Pi_s^2\Pi_{dd}\}.$$

We also note that

$$1 - \gamma^2 = 1 - [(\Pi_s - \Pi_d)/(\Pi_s + \Pi_d)]^2 = 4\Pi_s\Pi_d/(1 - \Pi_t)^2.$$

Thus,

$$\Pi_d^2\Pi_{ss} - 2\Pi_s\Pi_d\Pi_{sd} + \Pi_d^2\Pi_{dd} \leq (1 - \gamma^2)(1 - \Pi_t)^3/8.$$

We conclude with some remarks on the meaning of the assumption that (3.5.5), the asymptotic variance, is not zero. It is clear from (A7.5) and the following lines that (3.5.5) = 0 if and only if

$$\Pi_d \mathcal{R}_{ab}^{(s)} - \Pi_s \mathcal{R}_{ab}^{(d)} = 0$$

for all (a, b) such that $\rho_{ab} \neq 0$. Multiplying through by $\rho_{ab}\mathcal{R}_{ab}^{(s)}$ or $\rho_{ab}\mathcal{R}_{ab}^{(d)}$ and summing, we find that, if (3.5.5) = 0, then

$$\Pi_d\Pi_{ss} = \Pi_s\Pi_{sd}, \qquad \Pi_s\Pi_{dd} = \Pi_d\Pi_{sd}.$$

The converse is immediate. We may write these two statements as

$$\Pi_{ss}/\Pi_s = \Pi_{sd}/\Pi_d, \quad \text{and} \quad \Pi_{dd}/\Pi_d = \Pi_{sd}/\Pi_s,$$

providing that Π_s and Π_d are different from zero. From these we see that (3.5.5) is zero if and only if, taking three individuals, 1, 2, 3, from the population at random,

$$\Pr\left\{\begin{matrix}\text{1 and 3 have “positive”} \\ \text{sign relation}\end{matrix} \middle| \begin{matrix}\text{1 and 2 have “positive”} \\ \text{sign relation}\end{matrix}\right\}$$
$$= \Pr\left\{\begin{matrix}\text{1 and 3 have “positive”} \\ \text{sign relation}\end{matrix} \middle| \begin{matrix}\text{1 and 2 have “negative”} \\ \text{sign relation}\end{matrix}\right\}$$

and

$$\Pr\left\{\begin{matrix}\text{1 and 3 have “negative”} \\ \text{sign relation}\end{matrix} \middle| \begin{matrix}\text{1 and 2 have “negative”} \\ \text{sign relation}\end{matrix}\right\}$$
$$= \Pr\left\{\begin{matrix}\text{1 and 3 have “negative”} \\ \text{sign relation}\end{matrix} \middle| \begin{matrix}\text{1 and 2 have “positive”} \\ \text{sign relation}\end{matrix}\right\}.$$

We suggest that this is an unlikely state of affairs in most applications. For example, if the four corner cells of the cross classification have positive probabilities ($\rho_{11}, \rho_{1\beta}, \rho_{\alpha 1}, \rho_{\alpha\beta} > 0$), then (3.5.5) must be positive.

Reprinted from the JOURNAL OF THE AMERICAN STATISTICAL ASSOCIATION
June 1972, Volume 67, Number 338
Theory & Methods Section

MEASURES OF ASSOCIATION FOR CROSS CLASSIFICATIONS. IV: SIMPLIFICATION OF ASYMPTOTIC VARIANCES

LEO A. GOODMAN and WILLIAM H. KRUSKAL*

The asymptotic sampling theory discussed in our 1963 article [3] for measures of association presented in earlier articles [1, 2] turns on the derivation of asymptotic variances that may be complex and tedious in specific cases. In the present article, we simplify and unify these derivations by exploiting the expression of measures of association as ratios. Comments on the use of asymptotic variances, and on a trap in their calculation, are also given.

1. INTRODUCTION AND SUMMARY

In our 1963 article [3], we discussed asymptotic sampling theory for some of the measures of association presented in our earlier articles [1, 2]. It would have been impractical to present in [3] asymptotic results for many measures under many sampling methods, so we gave results only for some of the more important cases, together with general methods so that others might more readily do their own asymptotics.

In the present article, we present a more unified way to derive asymptotic variances (which form the nub of the asymptotic theory) for the following two sampling methods:

a. Multinomial sampling over the entire two-way cross classification;
b. Independent multinomial sampling in the rows when total row proportions are known; that is, stratified sampling in the rows.

---

* Leo A. Goodman is Charles L. Hutchinson Distinguished Service Professor of Statistics and Sociology, University of Chicago, and research associate at the Population Research Center of the University; William H. Kruskal is professor of statistics and chairman, Department of Statistics, University of Chicago, 1118 East 58th Street, Chicago, Ill. 50637. This research was supported in part by National Science Foundation Research Grants Nos. NSF GS 2818 and NSF GP 16071. Some of the second author's work on the article was done while he was a Fellow at the Center for Advanced Studies in the Behavioral Sciences and holding a National Science Foundation Senior Post-doctoral Fellowship. The authors are grateful to Stephen Fienberg (University of Chicago) and to Robert Somers (University of California, Berkeley) for helpful suggestions.

We then apply this more unified view to several measures of association and obtain fresh formulas for asymptotic variances. Further to illustrate the method, we rederive a few of the asymptotic variances in [3]. We also take this opportunity to correct (4.4.3) of [3]. Use of the asymptotic variances in practice is also discussed, and finally we give a cautionary note about a trap when working out asymptotic variances.

The basic notion of this article is to exploit the expression of most measures of association as ratios, and to do a portion of the manipulations towards asymptotic variances in advance.

2. MULTINOMIAL SAMPLING OVER THE ENTIRE TWO-WAY CROSS CLASSIFICATION

2.1. The General Case

As before, let ρ_{ab} be the probability for the a, b cell of the cross classification, and let R_{ab} be the corresponding observed proportion. The ranges are $a=1, 2, \cdots, \alpha$ and $b=1, 2, \cdots, \beta$. The ρ_{ab} are unrestricted, except of course that they are non-negative and add to one.

Nearly all the measures of association we presented in our earlier articles were written as ratios. We exploit that structure and consider a generic measure in the form $\zeta=\nu/\delta$, where ν and δ are mnemonically chosen to stand for νumerator and δenominator; ζ is the generic measure of association, although it could be any ratio of two functions, ν and δ, of the ρ_{ab}'s. (We assume that $\delta\neq 0$ and that ν and δ are differentiable at the needed values of their arguments.)

In this section we assume a multinomial sample of size n over the entire $\alpha\times\beta$ cross classification, so that the R_{ab}'s are the maximum likelihood estimators of the ρ_{ab}'s. It is well known that the covariance between $\sqrt{n}(R_{ab}-\rho_{ab})$ and $\sqrt{n}(R_{a'b'}-\rho_{a'b'})$ is

$$\overset{K}{\delta}_{aa'}\overset{K}{\delta}_{bb'}\rho_{ab} - \rho_{ab}\rho_{a'b'}, \tag{2.1}$$

where the δ^K here is the Kronecker delta; the K superscript is used to avoid confusion with the denominator delta above.

The maximum likelihood estimators of ζ, ν, and δ are the same functions of the R_{ab}'s as ζ, ν, and δ are of the ρ_{ab}'s; we call these Z, N, and D, respectively, so that $Z=N/D$, and our concern is with the variance of the asymptotic normal distribution of $\sqrt{n}(Z-\zeta)$. (See Section 4 for comments on the possibility that $D=0$.) Let us define

$$\nu'_{ab} = \partial\nu/\partial\rho_{ab}, \qquad \delta'_{ab} = \partial\delta/\partial\rho_{ab}, \tag{2.2}$$

since these partial derivatives enter frequently. It is convenient also to define

$$\begin{aligned}\phi_{ab} &= \nu\delta'_{ab} - \delta\nu'_{ab},\\ \bar{\phi} &= \textstyle\sum_{a,b} \rho_{ab}\phi_{ab}.\end{aligned} \tag{2.3}$$

Note that $\bar{\phi}$ is a weighted average of the ϕ_{ab}'s.

Then by applying the delta method and the supplementary tools presented in [3], it is easily calculated that the asymptotic variance of $\sqrt{n}(Z-\zeta)$ is

$$\begin{aligned}\sigma^2 &= \frac{1}{\delta^4}\textstyle\sum_{a,b} \rho_{ab}(\phi_{ab} - \bar{\phi})^2\\ &= \frac{1}{\delta^4}\left\{\textstyle\sum_{a,b} \rho_{ab}\phi_{a,b}^2 - \bar{\phi}^2\right\}.\end{aligned} \tag{2.4}$$

This is a simple general form for the asymptotic variance. We proceed to three examples, the first a reprise from [3] and the second two new.

2.2. The Measure of Association Gamma

In [1] we discussed a measure of association called γ that might be appropriate when both polytomies of the cross classification are ordered. For convenience we recapitulate the definition of γ; its interpretation is given in [1]. In the present setting, γ is most readily defined by a short chain of definitions,

$$\begin{aligned}&\nu = \Pi_s - \Pi_d, && \delta = \Pi_s + \Pi_d,\\ &\Pi_s = 2\textstyle\sum_{a,b} \rho_{ab}\Pi_{\mathrm{I};ab}, && \Pi_d = 2\textstyle\sum_{a,b} \rho_{ab}\Pi_{\mathrm{IV};ab},\\ &\Pi_{\mathrm{I};ab} = \textstyle\sum_{a'>a}\sum_{b'>b} \rho_{a'b'}, && \Pi_{\mathrm{IV};ab} = \textstyle\sum_{a'>a}\sum_{b'<b} \rho_{a'b'}.\end{aligned} \tag{2.5}$$

The Roman numeral subscripts are mnemonics for the "quadrants" relative to (a, b). (Quadrants here refer to the conventional Cartesian system with $a'-a$ corresponding to the horizontal axis and $b'-b$ to the vertical.)

To calculate the derivatives it helps to look at an example: what is the derivative of Π_s with respect to ρ_{22}? Think of Π_s as two times

$$\rho_{11}\Pi_{\mathrm{I};11} + \rho_{12}\Pi_{\mathrm{I};12} + \rho_{21}\Pi_{\mathrm{I};21} + \rho_{22}\Pi_{\mathrm{I};22} + \cdots.$$

Then one term of the derivative comes from the fourth summand above: $2\Pi_{I;22}$. Another term comes from the first summand: $2\rho_{11}$. No other summands contribute terms, and hence the desired derivative is $2(\Pi_{I;22}+\rho_{11})$. In general, the derivative of Π_s with respect to ρ_{ab} is $2(\Pi_{I;ab}+\Pi_{III;ab})$, where

$$\Pi_{III;ab} = \sum_{a'<a} \sum_{b'<b} \rho_{a'b'}$$

in accord with our mnemonic.[1] Similarly, the derivative of Π_d with respect to ρ_{ab} is $2(\Pi_{IV;ab}+\Pi_{II;ab})$, and hence

$$\nu'_{ab} = 2(\Pi_{I;ab} - \Pi_{II;ab} + \Pi_{III;ab} - \Pi_{IV;ab}),$$

$$\delta'_{ab} = 2(\Pi_{I;ab} + \Pi_{II;ab} + \Pi_{III;ab} + \Pi_{IV;ab}).$$

In terms of the notation of A7 of [3, p. 362],

$$\nu'_{ab} = 2(\mathcal{R}_{ab}^{(s)} - \mathcal{R}_{ab}^{(d)}), \qquad \delta'_{ab} = 2(\mathcal{R}_{ab}^{(s)} + \mathcal{R}_{ab}^{(d)}).$$

Thus the general expression of (2.3) becomes

$$\begin{aligned}\phi_{ab} &= 2(\Pi_s - \Pi_d)(\mathcal{R}_{ab}^{(s)} + \mathcal{R}_{ab}^{(d)}) \\ &\quad - 2(\Pi_s + \Pi_d)(\mathcal{R}_{ab}^{(s)} - \mathcal{R}_{ab}^{(d)}) \\ &= 4[\Pi_s \mathcal{R}_{ab}^{(d)} - \Pi_d \mathcal{R}_{ab}^{(s)}], \end{aligned} \tag{2.6}$$

$$\bar{\phi} = 4 \sum_{a,b} \rho_{ab}[\Pi_s \mathcal{R}_{ab}^{(d)} - \Pi_d \mathcal{R}_{ab}^{(s)}] = 0$$

since $\sum \rho_{ab}\mathcal{R}_{ab}^{(d)} = \Pi_d$ and $\sum \rho_{ab}\mathcal{R}_{ab}^{(s)} = \Pi_s$.

Hence, if G is the sample analog of γ, the asymptotic variance of $\sqrt{n}(G-\gamma)$ is

$$\frac{16}{(\Pi_s + \Pi_d)^4} \sum_{a,b} \rho_{ab}[\Pi_s \mathcal{R}_{ab}^{(d)} - \Pi_d \mathcal{R}_{ab}^{(s)}]^2. \tag{2.7}$$

Except for minor notational differences, and the carrying out of the square, (2.7) is the same as (3.5.5) of [3]. (To see the equivalence, note that $\Pi_s+\Pi_d=1-\Pi_t$, and refer to the manipulations at the bottom of p. 362 of [3].) For some purposes the present form may be simpler than the form in [3].

---

[1] One might at first be concerned that these calculations of derivatives do not take into account that the sum of all the ρ_{ab} is one. That does not, however, introduce any difficulty so long as we use the correct covariance structure, which necessarily reflects linear restrictions like $\sum_{a,b} \rho_{ab}=1$, or its sample analogue $\sum_{a,b} R_{ab}=1$. There is a related possible difficulty, however, that is discussed in Section 6.

2.3. Somers' Asymmetrical Δ_{ba}

Somers [5] has introduced an asymmetrical modification[2] of γ,

$$\Delta_{ba} = \frac{\Pi_s - \Pi_d}{1 - \sum \rho_{a\cdot}^2}, \tag{2.8}$$

together with its mate Δ_{ab}, with the denominator replaced by $1 - \sum \rho_{\cdot b}^2$. (Following our past convention, a dot replacing a subsubscript means summation over the replaced subscript, e.g., $\rho_{a\cdot} = \sum_b \rho_{ab}$). The denominator of Δ_{ba} is the probability that two independently chosen units from a population governed by the $\{\rho_{ab}\}$ cross classification probabilities do not lie in the same row (are not tied in a).

An interpretation of Δ_{ba} may be given in terms of two such independently chosen units; let us call their (random) row and column numbers $(\boldsymbol{a}, \boldsymbol{b})$ and $(\boldsymbol{a}', \boldsymbol{b}')$, respectively, and say that the units are weakly concordant when $(\boldsymbol{a}-\boldsymbol{a}')(\boldsymbol{b}-\boldsymbol{b}')\geq 0$, i.e., when the order of the two columns is the same as that of the two rows or when there is a tie. Similarly we may define weak discordance as $(\boldsymbol{a}-\boldsymbol{a}')\cdot(\boldsymbol{b}-\boldsymbol{b}')\leq 0$. Then the conditional probability of weak concordance, given that there is a difference between the rows, is $[\Pi_s + \sum_b \rho_{\cdot b}^2 - \sum_{a,b} \rho_{ab}^2]/[1 - \sum_a \rho_{a\cdot}^2]$, and the conditional probability of weak discordance, given $\boldsymbol{a}\neq\boldsymbol{a}'$, is $[\Pi_d + \sum_b \rho_{\cdot b}^2 - \sum_{a,b} \rho_{ab}^2]/[1 - \sum_a \rho_{a\cdot}^2]$. Hence Δ_{ba} is the difference between these two conditional probabilities.

It is sometimes useful to write the denominator of Δ_{ba} in the form $\Pi_s + \Pi_d + \sum_{a,b} \rho_{ab}(\rho_{\cdot b} - \rho_{ab})$, where the third summand, which is equal to $\sum_b \rho_{\cdot b}^2 - \sum_{a,b} \rho_{ab}^2$, may be thought of as the probability that two independent random units are tied in column but not in row.

Some important properties of Δ_{ba} are:

1. Δ_{ba} is indeterminate if and only if the population is concentrated in a single *row*;
2. Δ_{ba} is 1 if and only if both Π_d and $\sum_{a,b} \rho_{ab}(\rho_{\cdot b} - \rho_{ab})$ are zero. The second condition says that each column has at most one non-zero cell: hence, after removing all-zero columns, Δ_{ba} is 1 if and only if the non-zero cells descend in staircase fashion, perhaps with treads of unequal width. A similar interpretation holds for $\Delta_{ba} = -1$;
3. Δ_{ba} is 0 in the case of independence, but the converse need not hold except in the 2×2 case.

---

[2] Somers used the symbol d_{ba}. We have changed this to Δ_{ba} in order to use d_{ba} for the sample analogue.

The numerator ν is the same for Δ_{ba} as for γ, and the denominator δ is $1-\sum \rho_{a\cdot\cdot}^2$. Hence ν'_{ab} is the same as for γ and, more simply, $\delta'_{ab}=-2\rho_{a\cdot\cdot}$ It follows that

$$\begin{aligned} \phi_{ab} &= -2\rho_{a\cdot}\nu - 2\delta(\mathcal{R}_{ab}^{(s)} - \mathcal{R}_{ab}^{(d)}), \\ \bar{\phi} &= -2\nu \textstyle\sum_a \rho_{a\cdot}^2 - 2\delta(\Pi_s - \Pi_d) = -2\nu, \end{aligned} \qquad (2.9)$$

and that

$$\phi_{ab} - \bar{\phi} = 2\nu(1 - \rho_{a\cdot}) - 2\delta(\mathcal{R}_{ab}^{(s)} - \mathcal{R}_{ab}^{(d)}). \qquad (2.10)$$

Letting d_{ba} denote the sample analogue to Δ_{ba}, the desired asymptotic variance for $\sqrt{n}(d_{ba}-\Delta_{ba})$ is, therefore,

$$\frac{4}{\delta^4} \textstyle\sum_{a,b} \rho_{ab}\left[\nu(1 - \rho_{a\cdot}) - \delta(\mathcal{R}_{ab}^{(s)} - \mathcal{R}_{ab}^{(d)})\right]^2. \qquad (2.11)$$

So far as we know, this result is newly published.

2.4. The Measure of Association τ_b

In [1] we presented an asymmetrical measure of association, τ_b, based on a notion—suggested to us by W. A. Wallis—of reconstructing as best one can the cross classification probabilities. Interpretative details are given in Section 9 of [1]. In the present setting, it is easier to work with $1-\tau_b$ (which will not affect the asymptotic variance) and to express $1-\tau_b$ in terms of its numerator and denominator,

$$\nu = 1 - \textstyle\sum_{a,b} (\rho_{ab}^2/\rho_{a\cdot}), \qquad \delta = 1 - \textstyle\sum_b \rho_{\cdot b}^2. \qquad (2.12)$$

The maximum likelihood estimator of $1-\tau_b$ is

$$1 - t_b = \frac{1 - \sum_{a,b} (R_{ab}^2/R_{a\cdot})}{1 - \sum_b R_{\cdot b}^2},$$

and we want to find the asymptotic variance of $\sqrt{n}(t_b-\tau_b)$, which is the same as that of $\sqrt{n}[(1-t_b)-(1-\tau_b)]$. Following our general prescription, we find

$$\begin{aligned} \delta'_{ab} &= -2\rho_{\cdot b} \\ \nu'_{ab} &= -\textstyle\sum_{a',b'} \frac{1}{\rho_{a'\cdot}^2}\left\{2\rho_{a\cdot}\rho_{ab}\delta_{aa'}^K\delta_{bb'}^K - \rho_{a'b'}^2\delta_{aa'}^K\right\} \\ &= -2\rho_{ab}/\rho_{a\cdot} + \textstyle\sum_{b'} (\rho_{ab'}^2/\rho_{a\cdot}^2), \end{aligned}$$

where δ^K is the Kronecker delta. Thus

$$\phi_{ab} = -2\nu\rho_{\cdot b} + 2\delta\rho_{ab}/\rho_{a\cdot} - \delta \sum_{b'} (\rho_{ab'}^2/\rho_{a\cdot}^2),$$

$$\begin{aligned}
\bar{\phi} &= -2\nu \sum_b \rho_{\cdot b}^2 + \delta \sum_{a,b} (\rho_{ab}^2/\rho_{a\cdot}) \\
&= -2\nu(1-\delta) + \delta(1-\nu) \\
&= -2\nu + \nu\delta + \delta \\
&= (1+\nu-2\tau_b)\delta \\
&= [\sum_{a,b} (\rho_{ab}^2/\rho_{a\cdot})][1 + \sum_b \rho_{\cdot b}^2] - 2\sum_b \rho_{\cdot b}^2.
\end{aligned} \tag{2.13}$$

(We have expressed $\bar{\phi}$ in several different ways for convenience of reference.) Hence the asymptotic variance of $\sqrt{n}(t_b - \tau_b)$ under full $\alpha \times \beta$ multinomial sampling is

$$\frac{1}{\delta^4} \sum_{a,b} \rho_{ab}(\phi_{ab} - \bar{\phi})^2, \tag{2.14}$$

where δ, ϕ_{ab}, and $\bar{\phi}$ are defined above.

This result is new; we did not discuss the distribution of t_b under full multinomial sampling in [3].

3. INDEPENDENT MULTINOMIAL SAMPLING IN THE ROWS

3.1. The General Case

In Section 4 of [3] we dealt with a stratified sampling method that may sometimes arise. In this method there are separate, independent multinomial samples within each row of the cross classification; further, the sample size in row a is $n_{a\cdot} = n\omega_a$, where the ω_a's are supposed known, positive, and summing to one. (In practice, $n\omega_a$ will not in general be an integer, so one would take $n_{a\cdot}$ as that integer closest to $n\omega_a$. For purposes of asymptotic theory, we need only assume that the $n_{a\cdot}/(n\omega_a)$ have the limit one as n grows large.) We also assume that the $\rho_{a\cdot}$'s are known and positive.

To be specific, we treat separate sampling within rows; if there is separate sampling within columns, one need only interchange the roles of rows and columns. The sampling method under discussion may arise either from stratification of a single population, or in comparing several different populations.

If $\omega_a = \rho_{a\cdot}$, that is if $n_{a\cdot}$ is proportional to $\rho_{a\cdot}$, matters simplify a bit because then R_{ab} is still the maximum likelihood estimator of ρ_{ab}, just as in the case of full multinomial sampling. It was for this reason that we restricted ourselves in [3] to the proportional sampling rate case. The general case, however, is not essentially harder—it requires only carrying multiplicative factors along—and, to save space, we deal with it directly. (This remark qualifies a possibly misleading statement in the last paragraph of Section 4.4 of [3].)

It is convenient to deal with conditional row probabilities, and accordingly we write

$$\tilde{\rho}_{ab} = \rho_{ab}/\rho_{a\cdot}, \qquad \tilde{R}_{ab} = R_{ab}/R_{a\cdot}$$

for all relevant values of a and b. (For asymptotic purposes, there is no problem about the positiveness of $R_{a\cdot}$, since we have assumed $\omega_a > 0$ and $n_{a\cdot}$ is within 1 of $n\omega_a$, but there may be a problem with zero $R_{a\cdot}$ for small finite sample sizes.)

The maximum likelihood estimator of $\tilde{\rho}_{ab}$ is clearly $\tilde{R}_{ab}$; hence that of ρ_{ab} is $(\rho_{a\cdot}/R_{a\cdot})R_{ab}$, which is almost $(\rho_{a\cdot}/\omega_a)R_{ab}$. Further, the covariance between $\sqrt{n}(\tilde{R}_{ab} - \tilde{\rho}_{ab})$ and $\sqrt{n}(\tilde{R}_{a'b'} - \tilde{\rho}_{a'b'})$ is readily seen to be

$$\frac{1}{\omega_a} \overset{K}{\delta}_{aa'} \left[\overset{K}{\delta}_{bb'} \tilde{\rho}_{ab} - \tilde{\rho}_{ab}\tilde{\rho}_{ab'}\right]. \tag{3.1}$$

Note the $1/\omega_a$ factor, and the first Kronecker delta outside the brackets, because of independence among the multinomials.

As before, we let $\zeta = \nu/\delta$ be a generic measure of association, but now we regard ζ, ν, and δ as functions of the $\tilde{\rho}_{ab}$'s (and, of course, of the known $\rho_{a\cdot}$'s). Then Z, N, D $(Z = N/D)$ are the corresponding sample quantities, obtained by replacing $\tilde{\rho}_{ab}$ with $\tilde{R}_{ab}$ to obtain maximum likelihood estimators. The $\rho_{a\cdot}$'s stay unchanged since they are known constants.

Next, let ν^*_{ab} and δ^*_{ab} be the partial derivatives of ν and δ, respectively, with respect to $\tilde{\rho}_{ab}$. We use asterisks instead of primes to avoid confusion about the argument of differentiation. (Nonetheless, we record the relationships $\nu^*_{ab} = \rho_{a\cdot}\nu'_{ab}$ and $\delta^*_{ab} = \rho_{a\cdot}\delta'_{ab}$.)

As in Section 2, we introduce

$$\overset{+}{\phi}_{ab} = \nu \overset{*}{\delta}_{ab} - \delta \overset{*}{\nu}_{ab}, \qquad \bar{\phi}^{+}_{a} = \textstyle\sum_b \tilde{\rho}_{ab} \overset{+}{\phi}_{ab}, \tag{3.2}$$

where $\bar{\phi}^{+}_{a}$ is a weighted average of the ϕ^{+}_{ab}'s in row a only. (Some symbolic distinction from the notation of (2.3) is necessary, and the + superscript is suggestive of fixed row marginals.) The methods of [3] then show easily that the asymptotic variance of $\sqrt{n}(Z-\zeta)$ is

$$\frac{1}{\delta^4}\sum_a \frac{1}{\omega_a}\sum_b \tilde{p}_{ab}(\overset{+}{\phi}_{ab} - \overset{+}{\bar{\phi}}_a)^2$$

$$= \frac{1}{\delta^4}\sum_a \frac{1}{\omega_a}\left[\sum_b \tilde{p}_{ab}\overset{+2}{\phi}_{ab} - \overset{+2}{\bar{\phi}}_a\right]. \tag{3.3}$$

In short, we obtain here a linear combination of the within-row variabilities of the $\overset{+}{\phi}_{ab}$'s (weighted by the $\tilde{p}_{ab}$'s), rather than the overall variability obtained in Section 2.

This is a simple general form for the asymptotic variance under independent sampling within rows. We illustrate it next with two examples, the first a generalization of Section 4.2 of [3], and the second a generalization and correction of formula (4.4.3) in [3].

3.2. The Measure of Association λ_b

In [1], we presented an asymmetrical measure of association, λ_b, based on the concept of optimal prediction. Interpretative details are given in Section 5.1 of [1]. In the present setting, it is easier to work with $1-\lambda_b$ (which will not affect the asymptotic variance) and to express $1-\lambda_b$ in terms of its numerator and denominator,

$$\nu = 1 - \sum_a \rho_{am}, \qquad \delta = 1 - \rho_{\cdot m}, \tag{3.4}$$

where ρ_{am} is the maximum of $\rho_{a1}, \cdots, \rho_{a\beta}$, and $\rho_{\cdot m}$ is the maximum of $\rho_{\cdot 1}, \cdots, \rho_{\cdot \beta}$. We shall assume, as in [3] (see Sec. 3.1 there), that ρ_{am} equals exactly one of the ρ_{ab}, say $\rho_{ab(a)}$, and that $\rho_{\cdot m}$ equals exactly one of the $\rho_{\cdot b}$, say $\rho_{\cdot b(\cdot)}$ $(b=1, \cdots, \beta)$. (Note about symbolism: In [3] we used $\rho_{a(\cdot m)}$ for what is called $\rho_{ab(\cdot)}$ here, where $b_{(\cdot)}$ is the value of the column subscript index maximizing $\rho_{\cdot 1}, \cdots, \rho_{\cdot\beta}$; and we did not in [3] need a symbol for what we now call $b_{(a)}$, the value of the column subscript index maximizing $\rho_{a1}, \cdots, \rho_{a\beta}$.)

The maximum likelihood estimator of $1-\lambda_b$ is

$$1 - L_b = \frac{1 - \sum_a (\rho_{a\cdot}\tilde{R}_{am})}{1 - \underset{b}{\mathrm{Max}} \sum_a (\rho_{a\cdot}\tilde{R}_{ab})},$$

and we want to find the asymptotic variance of

$\sqrt{n}(L_b-\lambda_b)$, which is the same as that of $\sqrt{n}[(1-L_b)-(1-\lambda_b)]$. To follow our general prescription,[3] first write

$$\nu = 1 - \textstyle\sum_a (\rho_{a\cdot}\tilde{\rho}_{am}), \qquad \delta = 1 - \underset{b}{\text{Max}} \textstyle\sum_a (\rho_{a\cdot}\tilde{\rho}_{ab}). \quad (3.5)$$

Recalling that the $\rho_{a\cdot}$'s are fixed, we differentiate with respect to $\tilde{\rho}_{ab}$ to find

$$\overset{*}{\nu}_{ab} = -\rho_{a\cdot}\overset{K}{\delta}_{bb(a)}, \qquad \overset{*}{\delta}_{ab} = -\rho_{a\cdot}\overset{K}{\delta}_{bb(\cdot)}.$$

A schematic sketch of the cross classification will aid in seeing why these are the derivatives. It follows that

$$\begin{aligned}\overset{+}{\phi}_{ab} &= -\nu\rho_{a\cdot}\overset{K}{\delta}_{bb(\cdot)} + \delta\rho_{a\cdot}\overset{K}{\delta}_{bb(a)},\\ \overset{+}{\bar{\phi}}_{a} &= \delta\rho_{ab(a)} - \nu\rho_{ab(\cdot)}.\end{aligned} \qquad (3.6)$$

Hence the asymptotic variance of $\sqrt{n}(L_b-\lambda_b)$ under independent sampling within rows is, from the second form of (3.3),

$$\begin{aligned}&\frac{1}{\delta^4}\sum_a \frac{1}{\omega_a}\Big[\sum_b \tilde{\rho}_{ab}\Big\{\delta^2\rho_{a\cdot}^2\overset{K}{\delta}_{bb(a)} - 2\delta\nu\rho_{a\cdot}^2\overset{K}{\delta}_{bb(a)}\overset{K}{\delta}_{bb(\cdot)}\\ &\qquad + \nu^2\rho_{a\cdot}^2\overset{K}{\delta}_{bb(\cdot)}\Big\} - \overset{+}{\bar{\phi}}{}_a^{2}\Big]\\ &= \frac{1}{\delta^4}\Big\{\delta^2\sum_a \theta_a\rho_{ab(a)}(1-\tilde{\rho}_{ab(a)}) \qquad (3.7)\\ &\qquad - 2\delta\nu\Big[\sum{}^r(\theta_a\rho_{ab(a)}) - \sum_a \theta_a\rho_{ab(a)}\tilde{\rho}_{ab(\cdot)}\Big]\\ &\qquad + \nu^2\sum_a \theta_a\rho_{ab(\cdot)}(1-\tilde{\rho}_{ab(\cdot)})\Big\},\end{aligned}$$

where $\theta_a = \rho_{a\cdot}/\omega_a$ and $\sum^r$ denotes summation over those values of a for which $b_{(a)} = b_{(\cdot)}$. (This summation usage had been used in [3].)

The quantity in curly brackets to the right of the equality sign in (3.7) is, when all $\theta_a = 1$, exactly the same as (4.2.6) of [3], except for notation changes. Although these expressions appear rebarbative, they are often not difficult to use in specific cases; we illustrated the use of the sample analogue of (3.7) in Section 3.2 of [3].

[3] It might be feared that differentiation will lead to difficulty here because the maximum function is not everywhere differentiable. This problem is discussed in Sections A4 and A5 of [3]; there is no difficulty with the asymptotic theory under our assumptions.

3.3 The Measures of Association τ_b

We return to τ_b of Section 2.4, but now under independent sampling in the rows. It is convenient to work with $1-\tau_b$ and to express it as ν/δ, where

$$\nu = 1 - \sum_{a,b} \rho_{a\cdot}\tilde{\rho}_{ab}^{2}, \qquad \delta = 1 - \sum_b \Big(\sum_a \rho_{a\cdot}\tilde{\rho}_{ab}\Big)^2. \quad (3.8)$$

The maximum likelihood estimator of $1-\tau_b$ given in Section 2.4 should have $R_{a\cdot}$ replaced by $\rho_{a\cdot}$, since the $\rho_{a\cdot}$'s are known. Hence the estimator now is

$$\Big[1 - \sum_{a,b} \rho_{a\cdot}\tilde{R}_{ab}^{2}\Big]\Big/\Big[1 - \sum_b \Big(\sum_a \rho_{a\cdot}\tilde{R}_{ab}\Big)^2\Big].$$

Following our general prescription,

$$\overset{*}{\nu}_{ab} = -2\rho_{a\cdot}\tilde{\rho}_{ab} = -2\rho_{ab}, \qquad \overset{*}{\delta}_{ab} = -2\rho_{a\cdot}\rho_{\cdot b},$$

so that

$$\begin{aligned} \overset{+}{\phi}_{ab} &= 2\delta\rho_{ab} - 2\nu\rho_{a\cdot}\rho_{\cdot b} = 2\rho_{a\cdot}\big[\delta\tilde{\rho}_{ab} - \nu\rho_{\cdot b}\big], \\ \overset{+}{\bar{\phi}}_{a} &= 2\rho_{a\cdot}\Big[\delta \sum_b \tilde{\rho}_{ab}^{2} - \nu \sum_b \rho_{\cdot b}\tilde{\rho}_{ab}\Big]. \end{aligned} \quad (3.9)$$

It is helpful to let $\psi_{ab} = \delta\tilde{\rho}_{ab} - \nu\rho_{\cdot b}$, so that

$$\overset{+}{\phi}_{ab} = 2\rho_{a\cdot}\overset{+}{\psi}_{ab}, \qquad \overset{+}{\bar{\phi}}_{a} = 2\rho_{a\cdot}\overset{+}{\bar{\psi}}_{a},$$

where $\overset{+}{\bar{\psi}}_{a} = \sum_b \tilde{\rho}_{ab}\overset{+}{\psi}_{ab}$. In these terms, the desired asymptotic variance for independent sampling in rows is

$$\frac{4}{\delta^4} \sum_{a,b} \theta_a\rho_{ab}\big(\overset{+}{\psi}_{ab} - \overset{+}{\bar{\psi}}_{a}\big)^2. \quad (3.10)$$

When the sample sizes by rows are proportional to the $\rho_{a\cdot}$'s, i.e., when $\omega_a = \rho_{a\cdot}$ so that all $\theta_a = 1$, then (3.10) with "θ_a" deleted gives the asymptotic variance. Formula (4.4.3) of [3] purported to give that asymptotic variance, but in error; the second term of (4.4.3) of [3] is wrong.

4. USE OF THESE RESULTS IN PRACTICE

Probably the most common use of these results in practice (see Section 3.2 of [3]) is to treat $\sqrt{n}(Z-\zeta)/\hat{\sigma}$ as approximately unit-normal, where $\hat{\sigma}^2$ is a consistent estimator of the asymptotic variance σ^2. In the setting of our sequence of articles, $\hat{\sigma}^2$ is readily taken as the maximum likelihood estimator of σ^2, as follows.

Any σ^2 is a function of the ρ_{ab}'s, which may for convenience be written, perhaps in part, in terms of the $\tilde{\rho}_{ab}$'s. To find the maximum likelihood estimator of σ^2, make the replacements in the arguments of σ^2 as listed below. Recall that $R_{ab} = N_{ab}/n$, the proportion of all observations in the (a, b) cell, and that $\tilde{R}_{ab} = N_{ab}/n_{a\cdot}$, the proportion of observations in the (a, b) cell relative to row a.

Full multinomial sampling. (Sec. 2)

$$\rho_{ab} \rightarrow R_{ab} .$$

Independent sampling in rows. (Sec. 3)

$$\rho_{ab} \rightarrow R_{ab}(\rho_{a\cdot}/\omega_a) = R_{ab}\theta_a = \tilde{R}_{ab}\rho_{a\cdot}$$

$$\tilde{\rho}_{ab} \rightarrow \tilde{R}_{ab} .$$

In practice, $\hat{\sigma}$ may be zero; we discuss this problem in [3], e.g., in connection with γ on p. 324 of [3]. Provided that $\sigma > 0$, however, the probability that $\hat{\sigma} = 0$ approaches zero as n grows, so for large enough samples the $\hat{\sigma} = 0$ problem disappears. We have no analytic information about what "large enough" means, but we have encouraging evidence from the simulations reported in [3] and those of Rosenthal [4]. In the next section we consider the meaning of $\sigma = 0$ for the measures of association described earlier.

It can also happen in practice that $D = 0$, and then Z is undefined. This problem was discussed in [3, p. 320] for the case of λ_b. Since we assume throughout that $\delta \neq 0$, and since D converges to δ in probability, the $D = 0$ problem also vanishes as n gets large.

5. WHEN DOES $\sigma = 0$?

5.1. Full Multinomial Sampling

From (2.4), it is clear that, under full multinomial sampling, $\sigma = 0$ if and only if $\rho_{ab}(\phi_{ab} - \bar{\phi}) = 0$ for all a, b. What does this mean for the examples of Section 2?

Gamma. Criteria for $\sigma = 0$ in the case of G under full multinomial sampling were discussed in Section A7 of [3]. The basic condition given there may be rewritten, if $\Pi_d > 0$, as follows:

> If two individuals, 1 and 2, are drawn independently at random from the population, then, whenever $\rho_{ab} > 0$,
>
> $$\frac{\Pr\{2 \text{ is concordant with } 1 \mid 1 \text{ in } (a, b) \text{ cell}\}}{\Pr\{2 \text{ is discordant with } 1 \mid 1 \text{ in } (a, b) \text{ cell}\}}$$
>
> does not depend on the choice of (a, b), except that both numerator and denominator may be zero for some (a, b).

By interchanging numerator and denominator, a similar

condition may be written under the assumption $\Pi_s > 0$. By our general assumption, both Π_s and Π_d cannot be 0.

If either Π_s or Π_d is 0 (i.e., $\gamma = \pm 1$), then $\sigma = 0$. If at least one corner cell has positive probability, this becomes an equivalence: $\sigma = 0$ if and only if $\gamma = \pm 1$.

A family of cross classifications for which $\sigma = 0$ is the balanced cruciform family: all the probability is in a single row and column (neither of them borders), and there is equal probability in the two horizontal arms of the "cross" as well as equal probability in the two vertical limbs. A specific numerical example is

	.1			
.2	.4	.1		.1
	.05			
	.05			

where cells without numbers have zero probabilities. In such a balanced cruciform case, $\Pi_s = \Pi_d$ so $\gamma = 0$, and $\mathcal{R}_{ab}^{(s)} = \mathcal{R}_{ab}^{(d)}$ for every cell with $\rho_{ab} > 0$.

There are, however, other cross classifications with $\sigma = 0$, but for which γ is not -1, 0, or 1. All appear to be very special. For example, consider the 4×4 case in which there is probability 0.25 in cells (1,2), (2,1), (3,4), and (4,3); other cells have, of course, zero probability. Here $\Pi_s = .5$, $\Pi_d = .25$ so that $\gamma = 1/3$, yet it is easy to check that $\sigma = 0$ because the concordance-discordance ratio is 2 for the four cells with positive probability.

The measure Δ_{ba}. A necessary and sufficient condition that $\sigma = 0$ here is that, for all cells with $\rho_{ab} > 0$,

$$\mathcal{R}_{ab}^{(s)} - \mathcal{R}_{ab}^{(d)} = \Delta_{ba}(1 - \rho_{a\cdot}).$$

We do not have a neat characterization of these cases.

The measure τ_b. This was not discussed in [3] for full multinomial sampling. The major finding here is that if all $\rho_{ab} > 0$, then $\sigma = 0$ if and only if independence holds, i.e., if and only if $\rho_{ab} = \rho_{a\cdot}\rho_{\cdot b}$ for all (a, b). (Note that this implies $\tau_b = 0$.)

To see this, assume that all $\rho_{ab} > 0$, so that to say $\sigma = 0$ is to say (see (2.13)) that for all a, b_1, b_2

$$0 = \phi_{ab_1} - \phi_{ab_2} = -2\nu(\rho_{\cdot b_1} - \rho_{\cdot b_2}) + 2\delta(\tilde{\rho}_{ab_1} - \tilde{\rho}_{ab_2}).$$

Hence, taking the difference of this quantity between rows a_1 and a_2,

$$\tilde{\rho}_{a_1b_1} - \tilde{\rho}_{a_1b_2} = \tilde{\rho}_{a_2b_1} - \tilde{\rho}_{a_2b_2}.$$

Finally, add over b_2 and recall that $\tilde{\rho}_{a.} = 1$. It follows that $\tilde{\rho}_{a_1b_1} = \tilde{\rho}_{a_2b_1}$, or that $\rho_{a_1b_1} = \rho_{a_1}.\tilde{\rho}_{a_2b_1}$. Now average both sides over a_2 to obtain $\rho_{a_1b_1}$ as a product of a factor depending only on a_1 and another depending only on b_1. This shows independence. Conversely, if $\rho_{ab} = \rho_{a.} \cdot \rho_{.b}$ for all a, b, substitution shows immediately that $\delta = \nu$ and that all $\phi_{ab} = 0$.

If some ρ_{ab}'s are 0, we do not know a nice way to characterize $\sigma = 0$.

5.2. Independent Sampling in Rows

Here $\sigma = 0$ if and only if $\tilde{\rho}_{ab}(\phi_{ab}^+ - \bar{\phi}_a^+) = 0$ for all a, b. What does this mean for the examples of Section 3?

The measure λ_b. In [3, p. 315] we asserted that for λ_b, $\sigma = 0$ if and only if $\lambda_b = 0$ or 1, but we did not there give a proof. In our current notation and approach, a proof may be given relatively easily. We assume, without loss of generality, that all $\rho_{a.} > 0$; if $\rho_{a.} = 0$, just delete that row.

Recall first that

$$\lambda_b = 0 \quad \text{means } \rho_{ab(a)} = \rho_{ab(.)}, \quad \text{for all } a, \text{ or,}$$
$$\text{equivalently, } \sum^r \rho_{a.} = \sum \rho_{a.},$$
$$\lambda_b = 1 \quad \text{means } \sum \rho_{am} = \sum \rho_{ab(a)} = 1.$$

Now suppose that $\sigma = 0$, i.e., that $\tilde{\rho}_{ab}(\phi_{ab}^+ - \bar{\phi}_a^+) = 0$ for all a, b. In row a, look at the a, $b_{(a)}$ cell, for which $\tilde{\rho}_{ab(a)}$ must be positive. Thus, for all a,

$$\phi_{ab(a)}^+ - \bar{\phi}_a^+ = \delta(\rho_{a.} - \rho_{ab(a)}) - \nu(\rho_{a.}.\delta_{b(a)b(.)}^K - \rho_{ab(.)}) = 0.$$

Next, add over a, to obtain $\delta\nu - (\sum^r \rho_{a.} - \rho_{.m})\nu = 0$. Hence, either $\nu = 0$ (whence $1 = \sum \rho_{am}$ and $\lambda_b = 1$) or else

$$1 - \rho_{.m} - \sum^r \rho_{a.} + \rho_{.m} = 1 - \sum^r \rho_{a.} = 0.$$

If $1 = \sum^r \rho_{a.}$, then $\sum^r \rho_{a.} = \sum \rho_{a.}$, $\rho_{ab(a)} = \rho_{ab(.)}$ for all a, and $\lambda_b = 0$.

Conversely, if $\lambda_b = 1$, $1 - \lambda_b = 0$, $\nu = 0$, and $\rho_{a.} - \rho_{ab(a)} = 0$. Hence $\phi_{ab(a)}^+ - \bar{\phi}_a^+ = 0$. Similarly, if $\lambda_b = 0$, $1 - \lambda_b = 1$, $\delta = \nu$, and $\delta_{b(a)b(.)}^K = 1$ for all a. Hence

$$\phi_{ab(a)}^+ - \bar{\phi}_a^+ = \delta[\rho_{a.} - \rho_{ab(a)} - \rho_{a.} + \rho_{ab(.)}] = 0.$$

This completes the proof.

The measure τ_b. As in Section 5.1, the result here is that if all $\rho_{ab}>0$, $\sigma=0$ if and only if independence holds. The argument from independence to $\sigma=0$ is immediate; in the other direction, if all $\rho_{ab}>0$ and $\sigma=0$, then

$$\tfrac{1}{2}(\overset{+}{\phi}_{ab_1} - \overset{+}{\phi}_{ab_2}) = \delta(\tilde{\rho}_{ab_1} - \tilde{\rho}_{ab_2}) - \nu(\rho_{\cdot b_1} - \rho_{\cdot b_2}) = 0$$

for all a, b_1, b_2. The corresponding demonstration in Section 5.1 then applies.

6. CAUTIONARY NOTE ABOUT ASYMPTOTIC VARIANCES

In working out asymptotic variances of the above kind, there is a trap that stems from the singularity of the distributions, i.e., from relationships like $\sum_{a,b}\rho_{ab}=1$ or $\sum_b \rho_{ab}=\rho_{a\cdot}$. (See Footnote 1.) Because of these relationships, a given function of the ρ_{ab}'s may be expressed in a variety of ways, and sometimes one way is more convenient than another. Which way an expression is written makes no difference (except for convenience of computation) in the final asymptotic variance, *provided that* the same symbolic functional form is used throughout in finding derivatives. If not, incorrect results may be obtained.

We illustrate with a very simple case. Suppose that (X_{1n}, X_{2n}) form a sequence of pairs of random variables $(n=1, 2, 3, \cdots)$ such that $X_{1n}+X_{2n}=0$, and such that the pair $(\sqrt{n}(X_{1n}-2), \sqrt{n}(X_{2n}+2))$ has in the limit as n becomes large the (singular) bivariate normal distribution with means zero, variances 1, and covariance -1.

Note that we are treating the singularity consistently: first, $2+(-2)=0$; second, the asymptotic variance of $\sqrt{n}[(X_{1n}-2)+(X_{2n}+2)]$, which should be zero, is indeed $1-2+1=0$.

Now let the function of interest be $Y_n=X_{1n}^2$. Its derivative with respect to X_{1n} (evaluated at $X_{1n}=2$) is $2\times2=4$; the corresponding derivative with respect to X_{2n} is zero. Hence the asymptotic variance of $\sqrt{n}(Y_n-4)$ is $16(=(4)^2\times1)$.

But the function might just as well have been written $Y_n=X_{2n}^2$. The evaluated derivatives with respect to X_{1n}, X_{2n}, respectively, are 0 and -4. Hence the asymptotic variance of $\sqrt{n}(Y_n-4)$ is again 16.

A more interesting way of writing the function for illustrative purposes is $Y_n=\frac{1}{3}X_{1n}^2+\frac{2}{3}X_{2n}^2$. The evaluated derivatives now are $(\frac{2}{3})(2)=4/3$ and $(\frac{4}{3})(-2)=-8/3$, respectively. Hence the asymptotic variance is

$$\left(\frac{4}{3}\right)^2 - 2\left(\frac{4}{3}\right)\left(-\frac{8}{3}\right) + \left(-\frac{8}{3}\right)^2 = 16$$

as before. Thus, no matter how we choose to write the function, we get the same asymptotic variance, provided we remain faithful to the same symbolic form during the differentiation process.

If, however, we do not remain with one symbolic form, incorrect results may occur. In the above example, suppose that we write the function as X_{2n}^2 before getting the X_{1n} derivative, and as X_{1n}^2 before getting the X_{2n} derivative. Both evaluated derivatives will then be zero, and we will obtain the grossly wrong asymptotic variance of 0, instead of the foursquare correct value of 16.

REFERENCES

[1] Goodman, Leo A. and Kruskal, William H., "Measures of Association for Cross Classifications," *Journal of the American Statistical Association*, 49 (December 1954), 732–64.

[2] Goodman, Leo A. and Kruskal, William H., "Measures of Association for Cross Classifications, II: Further Discussion and References," *Journal of the American Statistical Association*, 54 (March 1959), 123–63.

[3] Goodman, Leo A. and Kruskal, William H., "Measures of Association for Cross Classifications, III: Approximate Sampling Theory," *Journal of the American Statistical Association*, 58 (June 1963), 310–64.

[4] Rosenthal, Irene, "Distribution of the Sample Version of the Measure of Association, Gamma," *Journal of the American Statistical Association*, 61 (June 1966), 440–53.

[5] Somers, Robert H., "A New Asymmetric Measure of Association for Ordinal Variables," *American Sociological Review*, 27 (December 1962), 799–811.

[*Received December 1970. Revised August 1971.*]